DUTCH

FROM RISING SUN TO THE RISE OF JIHAD
SIX DECADES OF SERVICE

KIM KIPLING

Copyright © 2022 by "Kim Kipling"

ALL RIGHTS RESERVED. This book or any portion thereof
may not be reproduced or used in any manner whatsoever without
the express written permission of the publisher except for the use
of brief quotations in a book review.

Printed in the United States of America

First Printing, 2022

ISBN 979-8-9856393-0-8 (softcover)
ISBN 979-8-9856393-1-5 (ebook)

SharpenedEdge@protonmail.com

THIS BOOK IS DEDICATED
TO ALL MY FRIENDS AND BROTHERS
WHO PAID THE HIGHEST PRICE,
MADE THE ULTIMATE SACRIFICE,
AND NEVER LEFT VIETNAM.

AND ALSO TO MY GRANDSONS,
WHO SACRIFICED THEIR TIME IN AFGHANISTAN AND IRAQ,
AND NOW DEDICATE THEIR LIVES TO THIS COUNTRY AS POLICE OFFICERS.
I AM A PROUD GRANDFATHER.

— DUTCH WIERENGA —

CONTENTS:

FOREWORD IX

NOTES ON REVIEW AND REDACTIONS XIII

PROLOGUE:
MAY, 1942. JAVA, DUTCH EAST INDIES XV

CHAPTER 1:
ECHOES OF COLONIALISM 1

CHAPTER 2:
EARLIEST BEGINNINGS 3

CHAPTER 3:
NEW FAMILY 7

CHAPTER 4:
A NEW LIFE AND VENTURE 9

CHAPTER 5:
RISING SUN 13

CHAPTER 6:
MIEL'S RELOCATION 15

CHAPTER 7:
RUN TO THE HILLS 19

CHAPTER 8:
THE JAPANESE INTERNMENT CAMP. 1942-1945 21

CHAPTER 9:
HELL WITHIN BARBED WIRE. INDONESIAN INTERNMENT, 1945-1946 27

CHAPTER 10:
GURKHA RESCUE 31

CHAPTER 11:
REBUILDING A LIFE. 1947-1952 35

CHAPTER 12:
FIRST COMBAT PATROLS. 1952 41

CHAPTER 13:
FALSE IMPRISONMENT, AND THE EFFICACY OF BRIBERY. 1953-1954 45

CHAPTER 14:
WORLD CRUISE! 1955 49

CHAPTER 15:
A SOLDIER, A BUREAUCRAT AND PLANS TO BECOME AN AMERICAN.
1955-1960 51

CHAPTER 16:
WELCOME TO AMERICA! 1960-1963 53

CHAPTER 17:
THE UNITED STATES ARMY. 1962-1966 57

CHAPTER 18:
FIRST MARRIAGE 61

CHAPTER 19:
CITIZENSHIP! 65

CHAPTER 20:
FIRST TOUR IN VIETNAM. JANUARY 1967 – SEPTEMBER 1968 67

CHAPTER 21:
SPECIAL FORCES SELECTION AND TRAINING 71

CHAPTER 22:
MACV/SOG 75

CHAPTER 23:
MACV/SOG COMMAND AND CONTROL NORTH.
SEPTEMBER 1969 – DECEMBER 1970 81

CHAPTER 24:
SOG MISSIONS 85

CHAPTER 25:
SOG PERSONALITIES 101

ILLUSTRATIONS 107

CHAPTER 26:
RT ANACONDA'S MISSIONS. 133

CHAPTER 27:
AMBUSH! THE SILVER STAR. 26 APRIL 1970 143

CHAPTER 28:
THE PURPLE HEART. 11 AUGUST 1970 153

CHAPTER 29:
SF RECON SCHOOL. JANUARY 1970 – APRIL 1974. 157

CHAPTER 30:
HALO SCHOOL. MAY 1974 – AUGUST 1977 161

CHAPTER 31:
ROTC AND SPECIAL FORCES TRAINING. SERE SCHOOL. RETIREMENT. 165

CHAPTER 32:
A NEW BEGINNING. THE CENTRAL INTELLIGENCE AGENCY. 1986-1999 173

CHAPTER 33:
THE CONTRAS. 1988-1991. 183

CHAPTER 34:
WEST AFRICA. 1994-1996. 189

CHAPTER 35:
THE FARM. 1996-1999 – PRESENT 191

CHAPTER 36:
CIA OPERATIONS TRAINING. 193

CHAPTER 37:
THE SOTC. 197

CHAPTER 38:
MRLN 201

CHAPTER 39:
WEAPONS TRAINING 205

CHAPTER 40:
THE OVERSEAS PERSONAL SECURITY COURSE 209

CHAPTER 41:
MARITIME OPERATIONS. 211

CHAPTER 42:
GROUND OPERATIONS. 213

CHAPTER 43:
AIRBORNE OPERATIONS. 217

CHAPTER 44:
RETIREMENT. SORT OF. 227

CHAPTER 45:
PASSING – AND A NEW JOY. 2006 – PRESENT 229

CHAPTER 46:
THE VERSTEEGH FAMILY, AND DUTCH TODAY 233

CHAPTER 47:
DUTCH'S SUMMARY. 239

ACKNOWLEDGEMENTS 243

ABOUT THE AUTHOR 245

INDEX 246

FOREWORD:

This book tells the story of a remarkable American patriot, who you most likely have never heard of. Master Sergeant Jan W. "Dutch" Wierenga, US Army, retired, was born in the Dutch East Indies (now known as Indonesia) in 1936. He was the child of a wealthy Dutch planter and a half-Indonesian mother who was descended from hereditary nobility. The upper-class family was quite comfortable...for a while. Then in February, 1942, everything changed when the Japanese Army invaded Indonesia. He and his mother and two sisters were imprisoned for two years in a Japanese internment camp where deprivation, disease and abuse were daily occurrences. Death was all around them, yet almost miraculously, they all survived to be liberated in 1944. Then things got worse: they were caught up in the violence of the Indonesian civil war. The little family was again imprisoned by the rebellious local government, in unspeakably harsh conditions which were far worse than those inflicted by the Japanese. This lasted for two more years, with all of them again by some miracle surviving intact. They were released in 1946, but were still in extreme danger from violent nationalist militia forces who were fighting to overthrow the Dutch colonial government, and who saw all Dutch citizens as the enemy. They were ultimately rescued under fire by British armed forces troops, flown to safety and reunited with other members of the family.

    After their release, Dutch returned to a cousin's rubber plantation, then under periodic attacks from agents of the Indonesian government's senior leadership who sought to drive off the foreign owners so that they could take the profits of the rubber production for themselves. By age sixteen, he was running combat patrols in the jungle areas of the plantation, armed with a Thompson submachine gun and fending off the frequent arson and sabotage attacks they were experiencing. Ultimately, the entire family was forced to flee Indonesia before they could again be imprisoned on false charges. Dutch made his way by cargo ship to Holland, by then an experienced, traveled and educated young man, courtesy of a Jesuit school run by an uncle. He joined the Dutch army, and completed an enlistment term while his application to emigrate to the United States was processed. He was approved, and traveled with several thousand other intending immigrants via steamship to New York. He traveled across America on the most spartan of trains, and moved in to work with the assistance of his sponsor, later pursuing a college education at a small community college in Marin County, California. He took a job with a Montgomery Wards department store.

But Dutch was determined to give back to his adoptive country, and he wanted to be a soldier. He enlisted in the U.S. Army in 1962, breezing through basic training. Never one to do anything halfway, he immediately volunteered to become a paratrooper. After a four-year assignment to Germany with the 82nd Airborne Division, he returned to the USA, became a US citizen, and in early 1967, Dutch found himself in Vietnam, in combat with the 101st Airborne Division. The Army then introduced him to the Special Forces, and in due time, he earned the coveted Green Beret. Then Dutch volunteered again; he joined the highly-classified Studies and Observations Group of the Military Assistance Command, Vietnam (MACV/SOG). This now-legendary unit of elite commandos ran some of the most dangerous, highly classified missions of the Vietnam War. As a Recon Team leader, Dutch was right in the thick of this pitched combat, earning numerous battle decorations including the Silver Star, four Bronze Stars and the Purple Heart. He later served as the senior NCO of the Special Forces Training School, and helped develop the techniques of High Altitude/Low Opening (HALO) parachuting as part of the very first HALO Committee. He retired from the U.S. Army after a storied career including three combat tours in Vietnam.

Then Dutch did something even more amazing: he volunteered again, joining the Central Intelligence Agency as a Paramilitary Operations Officer. For the next several years, he conducted clandestine missions all over the globe, in some of the worst conditions and most dangerous places imaginable. When time took its inevitable toll and field deployments were no longer possible for him, he continued to serve as a training instructor, and ultimately in a training support role. He finally fully retired in Summer 2022 – SIXTY YEARS after he first enlisted in the service of the United States of America.

In all, Dutch has served the United States in his several capacities for six decades. In a time before the Internet, so-called "reality" TV and a popular culture of shameless self-promotion, members of the Special Forces and officers of the CIA prided themselves on being "Quiet Professionals." They purposefully avoided the limelight, never calling attention to their deeds and heroism. "Let your actions and accomplishments speak for themselves" was a common motto. Dutch exemplified this quiet professionalism. He never sought attention. He never heralded his own actions, his successes, his enduring dedication to mission. He rarely even raised his voice. But when he spoke, experienced peers listened, and he enjoyed the universal respect and approbation of generations of SF soldiers and CIA officers. He served. He persevered. He gave much to America. This is his story.

NOTES ON REVIEW AND REDACTIONS:

As a retired CIA staff officer and current Contractor, I was required by law to submit the draft manuscript of this book to the Central Intelligence Agency for pre-publication review. To give credit where it is due, CIA turned around the manuscript draft in a very expeditious manner. However, they made significant redactions to the sections of the book which cover Dutch's CIA career. The redactions reflect the Agency's excessive and in many cases ridiculous reticence to disclose information which is either long out of date, or in many cases already disclosed in other writings which were not submitted for review. Nonetheless, I have edited the text to conform to these redactions as required. If any particular section seems overly vague concerning place locations, events, unit designations, etc., I beg your understanding. In most cases, you will probably be able to discern the specific data with a little reflection on the known history of the time or identical information available in the open record, with a high degree of probable accuracy. In the most egregious of these mandated omissions, I have left the text redacted exactly as CIA required, as examples of the thought process attendant to the review. CIA had one more requirement; that I include the following disclaimer:

"All statements of fact, opinion or analysis expressed are those of the author and do not reflect the official positions or views of the U.S. Government. Nothing in the contents should be construed as asserting or implying U.S. Government authentication of information or endorsement of the author's views."

Just in case you mistakenly thought that the author was the Director of the Central Intelligence Agency – I'm not.

"Kim Kipling"

PROLOGUE:

MAY, 1942.
JAVA, DUTCH EAST INDIES

The boy was playing quietly on the carpet in the bedroom, enjoying the sunshine streaming in through the open window. The sounds and scents of an upscale river resort hotel in the eastern Javanese hill country filled the morning air. Several servants went about their daily routines of cleaning, cooking, doing laundry, and maintaining the palatial house, as an extensive and productive rubber and coffee plantation in the countryside had made this family wealthy for generations, and their daily life reflected their wealth and status. Breakfast was over, and the boy's customary regimen of learning new words in several languages, reading, writing, simple numbers and the basic concepts of how things worked had not yet started for the day. He was only five years old, so did not yet attend the local school like his older siblings. He loved the time he could spend in the morning before his lessons, creating scenes of animals, adventures, and experiences in his mind. He wasn't really sure why things were changing so fast, and becoming so different. He knew that his parents and older brothers had been very worried about some people from a place called Japan who had recently come to Java. They had talked at length about these people, long into the night when he was supposed to be sleeping.

He could tell from their tone of voice that his father and brothers were both angry and fearful about the things these "bad men" from Japan were doing in Java and elsewhere. They talked about the "war" – a word he had not understood until he asked his mother what it meant. She had told him "war is when bad men come and try to hurt people, and those people have to fight them to make them stop. It is horrible. People kill each other. We must pray that it does not happen here." And not long after that night, his father and his eldest brother had gone away, to another part of Java. His mother had told him that they had joined the Dutch army, to fight against the bad men. They were soldiers now, and neither he nor his mother knew exactly where they were. He missed his father and brother very much.

Perceptive and bright for his age, the boy heard a new sound begin to come through the open window. It was the sound of rhythmic drumming, echoing up the valley from a small village near where the narrow road to the resort joined a much larger road leading to the nearby town of Magelang. This intersection was four miles away, but the sound of drumming carried well and could be clearly heard. As the steady rhythm continued, his imaginary world of toys was interrupted by the voice of his mother, calling his name with fear and urgency. "Jan! Come at once! The bad men are coming! We must leave now. Hurry!"

His nanny flew into the room and began packing some extra clothing for him into a bundle made of a bed sheet. Looking at the boy with dread-filled eyes, she softened and added a small toy or two to the load, then grabbed his hand and rushed him down the stairs to his mother's side. "But Mama, I don't want to leave! Why must we go? I want to stay here for my lessons!" But his mother grasped his hand tightly and simply told him "Quiet! We must hurry now. The bad men are coming, and they will hurt us if they catch us. We have to go up into the hills for a while, until they leave. Don't argue, just stay beside me, no matter what happens." And the boy, his mother and two older sisters, and several of the servants hurriedly left the house and began climbing the paths into the rain forest, higher into the hills above the resort. He carried his little bundle over his shoulder, as everyone else did with their own belongings. His mother and the others walked quickly, propelled by fear, and as he fell behind, his elder sister Patricia held his hand fiercely as she practically dragged him up the trail. He grew tired, and she carried him for a while, rushing onward, higher, further from home. The seasonal rains had not yet completely ceased and the pathway was muddy and slick, with some ruts where the carts passed slowing their progress. Branches occasionally whipped his face as they overhung the narrow trail edge. And frightened and tired, the boy began to cry. "Why, Patricia? Why are the bad men after us?" he asked his sister. "Because they are evil men, and they hate us. Because Mama has been helping some of our neighbors who are fighting to stop them. But we will not let them catch us. We will keep going, and only go home when they have left." And she gripped his hand even more tightly, shepherding her "little broeder" even as her own fear threatened to overwhelm her.

The boy saw clearly that his mother, sisters and the servants were all badly frightened. He saw that his sisters were also struggling to keep up with the frantic pace. And he hated the feeling of being frightened too. He thought of his father and elder brother, away being soldiers, fighting the bad men from Japan. And he realized that that was what soldiers did: fight the bad men to protect their families and homes. That seemed to him to be a very fine thing, a worthy thing. And he decided that one day, he wanted to be a soldier too. He also would fight to protect the things and people he loved. And in that moment, a warrior spirit was born. No one knew just how well and how long that spirit would fight, in so many places and so many ways. But a lifetime of enduring service began then and there, on that muddy trail in eastern Java.

CHAPTER 1:

ECHOES OF COLONIALISM

Twenty-first Century Holland is seen by most Americans as a quaint anachronism; it is a land of wooden shoes, windmills, tulips and dikes. Dutch beer is well-known, as are the seedier corners of the capital city, Amsterdam. The most enduring stereotype of the Dutch people is a stunning blonde, blue-eyed young woman wearing a country dress and apron while making Edam cheese. But the historical reality is somewhat more serious. Holland, or The Netherlands as its greater modern nation is now known, wasn't always a minor player in global affairs.

It has been said that, with regard to politics and in the most basic of terms, POWER can be defined as the ability to kill people and destroy things. INFLUENCE flows from and is directly proportional to POWER. And historically, Power had to be projected to grow significant Influence. With the exception of a few dominant land-based nations who could march major armies abroad, Power Projection generally meant Seapower. Holland had both power and influence in great measure during the 17th – 19th Centuries, largely as an outgrowth of formidable seapower, coupled with immense economic resources.

The Dutch people have always been fighters. The land itself had to be fought just to live and grow sustenance. Huge tracts had to be reclaimed from a cold and hostile sea, and a harsh winter climate made survival difficult. The recorded history of the land now known as the Netherlands is a cyclic pattern of invasion, resistance and conquest, encompassing bloody and cruelly savage conflict. Each in their times, Saxons, Vikings, Celts, Romans and Franks all fought over Dutch soil. This crucible of environmental duress and repeated warfare produced a particularly hardy, warlike, and fearless people, in whose bloodlines those formidable if sometimes malignant genes still course. In the 17th Century, a period known as the Dutch Golden Age, the Dutch Empire was one of the world's great seafaring and economic powers. Dutch shipbuilders built an astonishing fleet of 16,000 merchant ships and proportionally numerous naval vessels, all superbly designed and crafted given the available technologies of the time. Dutch seamen were justly world-famous. Expert celestial navigators and ocean sailors, they conned their excellent ships all around the world, claiming territories to form a global empire while trading, and in some cases, raiding. Dutch naval sailors and pirates were especially feared. Merciless, capable and aggressive fighters, they were formidable agents of national expansion and power projection. In their wake, the Dutch East India Company and Dutch West India Company established colonies and trading posts all over the world, from China to Japan, Taiwan, North America, South America and South Africa, and to the territories we now know as the modern nation of Indonesia. The spice trade made Holland a world power, rich beyond calculation, and it ultimately became a global center of finance and banking. (During the American Revolution, America's future second President, John Adams, was dispatched to Amsterdam to try and persuade the Dutch to lend considerable sums of money to our fledgling nation to finance its military expenses. Ever penurious and wary, the Dutch government only agreed to lend America money AFTER it had won the Revolutionary War.) Out of their national prosperity, Dutch science, art and military skills grew to be among the most acclaimed in the world. And the Netherlands retained considerable wealth and influence well into the Twentieth Century.

CHAPTER 2:

EARLIEST BEGINNINGS

Into this fading but not-yet-entirely-bygone world of Dutch economic and colonial influence, Emil Ignatius Versteegh was born on July 06th, 1901, in Holland. He was the youngest child of his family. The Versteeghs were extremely wealthy. Among other interests, they owned several thriving rubber and coffee plantations in one of Holland's remaining far-flung imperial possessions, the Dutch East Indies, in Eastern Java. At this time, rubber was an important strategic material; it was a necessary and critical component of many items of military, industrial and economic necessity. Rubber cannot be grown just anywhere, and it thrived in the fertile, humid jungles of Java. Local labor to grow, harvest and ship it was cheap and plentiful. So rubber production was extremely profitable. And as the precious latex sap flowed, so did the enormous profits it generated – straight into the Versteegh family bank accounts.

    Young Emil grew into a formidable young man. He was accustomed to deference and influence from birth. As heir to his family's tremendous wealth, he was given a classical European liberal education. He played football (what we know as "soccer" in America) very well, despite his heavy tobacco use. Emil would take advantage of the rare breaks in the game to drag down two cigarettes, and then retake the field to resume the near-constant running and occasional bursts of extremely kinetic activity which mark the world's most popular sport. He studied fencing, and grew to be sufficiently skilled in this traditional martial art as to be made an instructor while he was still in the university. Known as a stern, serious and capable young man, he entered full adulthood with a problem. Emil's three elder brothers were all Catholic priests. His sisters were nuns. Entering the Catholic clergy was simply a familial expectation; the Versteegh children were to be Catholic clerics, and that was all there was to it. But Emil felt no such calling, and he had no intention of taking up the priesthood. Once his resistance was expressed and his limits tested, his parents ultimately realized that the strong-willed young man could not be forced into seminary and made to take his priestly vows. So in the end, they had to do something to get him out of sight and to alleviate what they no doubt considered as the family's shame. Young Emil would be sent halfway around the world to live on and manage the family's thriving plantations in Eastern Java. In turn, Emil intuitively grasped the freedom and independence such an exile would allow him. And so it was that Emil emigrated to Eastern Java as a young man. He married a young Dutch woman from a good family, and she bore him two fine sons. Joseph, called "Joop" was born in 1921, followed by Emil, called "Miel" in 1927. Both boys shared their father's vitality and natural physical abilities. Miel in particular grew to be an incredible specimen of strength and coordination. He ultimately reached an adult height of 6'3". As a teenager, he would swim out into the nearby river and catch fish by hand. He loved to wrestle and could lift up the front end of a Jeep, engine and all. Both boys grew into vigorous adolescence in Java. Unfortunately, their mother did not share this trait of vibrant good health, and she died young, while Miel was still an infant.

 As the scion of a wealthy landowning family, Emil was a highly eligible widower. Given that he had an infant son, he was in need of a wife. And he found a likely candidate among the children of one of the oldest Dutch colonial families in Java. Carolina Arnoldina Wierenga was born on January 31st, 1906 in the ancient city of

Malang, Java. Known for its temperate climate and lovely scenery, Malang was a popular destination for European immigrants. Carolina's father, Joop Wierenga, was an educated, very wealthy and well-connected Dutch citizen. His father had in fact served as Palace Manager to the King of Indonesia in the early/mid-1800's. Reflective of Holland's strong Protestant historical dominance, the family were staunch Presbyterians. Unusually for the time, Carolina's mother was a native Javanese, but of royal lineage. Her bloodline directly traced back to the hereditary kings of Indonesia; no doubt her family's social and palace connections had made the racially-mixed match possible and encouraged the union for the mutual advantage of both families. The Wierengas lived in Malang and owned several profitable riverside plantations in Western Java, which contributed greatly to the family's continued prosperity. Carolina was a child of great wealth and privilege, but from an early age evinced a dark, moody personality. Her psychological instability was sufficiently severe that today she would probably be diagnosed as having bipolar disorder. Carolina would fly into fits of rage easily, and in her moments of pique would impulsively smash her expensive porcelain dolls which had been imported all the way from Holland. Chores and mundane domestic tasks were far beneath her and were never facets of her upbringing. Household servants saw to her every need. She was essentially raised as a princess, if not one formally invested with that court status. As she grew, Carolina's mixed blood manifested itself into the delicate profile, exotic beauty and exact facial symmetry often noted among young women of blended ethnicity. Lovely and rich, she had unlimited opportunities for courtship and eventual marriage into an equally prestigious family. When she grew to the proper age, the now-mature young woman was sent to Holland to complete an appropriate classical education for a lady of her time. And when she returned to Java, a stylish, cultured, wealthy and beautiful young woman, she caught Emil Versteegh's eye. In what seems to have been a case of mutual attraction, their courtship progressed into a full-blown love affair. Her father was not immediately convinced, and took some time before giving his permission for the marriage. He eventually relented and Carolina was able to accept Emil's proposal and to become his second wife, in 1927.

CHAPTER 3:

NEW FAMILY

Emil and his new bride Carolina set about expanding the Versteegh family immediately. His extensive coffee and rubber plantation was located adjacent to several other family-owned properties in the hills of Eastern Java. This area was remote and extremely bucolic; Carolina of course preferred the developed luxury and established society of the capital, Batavia (later to be renamed Jakarta.) So with the exception of a few visits to the plantation from time to time, she set up housekeeping in luxurious quarters in Batavia while Emil managed the affairs of the plantation day to day. He would visit the city occasionally (as Dutch remembers it: *"to smack me around a little bit from time to time"* in punishment for various juvenile misdeeds.) And in that manner, they built a pattern of married life featuring frequent geographic separations.

Carolina adjusted to her married life (as lived by a wealthy woman in the colonial Dutch East Indies,) with dozens of servants to see to her every daily necessity. As Dutch remembers it: *"My mother never cooked a day in her life."* Soon she was expecting a baby, and in 1928, a son, named Onno Verstccgh, joined the family. Onno was a stubborn and willful child, with a strong streak of mischief and resistance to authority running through his personality. This caused significant friction, as both his father and mother were strict, stern disciplinarians who brooked little nonsense in their progeny. In fact, Emil would occasionally become so frustrated with Onno's repeated transgressions and apparent lack of contrition that he would take him to the local orphanage and drop him off for a week or two, threatening to abandon him there for good! The tactic seems to have been ineffective. Dutch recalls: *"Onno made some of the best friends of his life from among the kids at the orphanage!"* And this trait would manifest itself throughout Onno's life.

In 1929, a daughter, Millie Versteegh, was born. Millie was a fair-haired, light-skinned baby girl, who blossomed in time into an attractive young woman who embodied the stereotypical image of the blonde, milky-complexioned Dutch maiden. She was, of all the children, her mother's favorite. In 1930, another daughter arrived. Patricia was sensitive, artistic and reserved. Her outward appearance reflected the East Asian genetic legacy of her native-born Grandmother. She had darker skin, hair and eyes than the typically fair Dutch colonial citizens of the area. She could, and in time did, pass as a native in public, and this may have saved the family's life when things were at their worst in future years. But for now, she was a healthy, if quiet, child.

Carolina then took a hiatus from childbearing. It seemed that the family was complete. But in 1936, there was one more arrival. A baby boy was born on the 31st of July, also bearing the unmistakable signs of his East Asian heritage. Dark, almond-shaped eyes and light brownish skin displayed the genes passed down from his Grandmother, the direct descendant of Indonesian kings. He was born with light blonde hair, but in time, this darkened to brown. And Emil and Carolina named him Jan Willem. In time, his American friends would come, predictably, to call him "Dutch."

CHAPTER 4:

A NEW LIFE AND VENTURE

Young Jan grew quickly, and initially lived an idyllic life in the lap of moneyed splendor. He was an intelligent, steady child. Servants attended to his every need. A nanny saw to his daily care; his mother was not the attentive, nurturing type, and he was her fourth child, so any sense of novelty in childbearing and rearing had long since worn off. When he was very small, Jan had to have his tonsils removed. His nanny went with him to the hospital and stayed with him constantly, sleeping underneath his hospital bed to care for his needs until he was discharged. In later years, Jan would waggishly remember: *"I was born with a silver spoon in my mouth....that quickly turned into a wooden spoon."* Given their history and economic status, this was a cultured and educated family. They learned and spoke five languages at home, quite naturally. Dutch and Bahasa Indonesian were spoken as native tongues. English was the language of education, while French and German were also learned as the cultural birthrights of sophisticated Europeans, to prepare them to compete successfully in that rarefied society. The family home was expansive, with many bedrooms, baths and other visible affectations of the high social class they occupied apparent. In the late 1930s in colonial Java, the family owned not one, but two automobiles.

In 1939, Emil decided that he had had enough of managing the family's remote ancestral rubber plantation. He longed for a different challenge and a more comfortable location. And he wanted to have the family all together instead of being geographically separated. So using the considerable resources at his disposal, he purchased a riverside hotel resort property in Kalibening, northwest of the city of Magelang, in Eastern Java. He moved the family there immediately, but retained ownership of the plantation, leaving it under the management of a caretaker. Kalibening means "clear river" and it lived up to its name. Although far from the capital in Jakarta, this was still a luxurious lifestyle for the family. The house had thirteen bedrooms, and a proportional number of bathrooms. Built sturdily, with very thick walls, it was a modest palace by the standards of the time. The resort also had numerous guest quarters, and no less than five swimming pools served by the clear water provided by the river. There was a whole compound in the back where the many servants lived. On Sundays, the family would go boating on the nearby lake for pleasure. At Christmas time, they would bring in an uprooted tree as a Christmas tree, and post a servant behind it to make sure the decorative candles they placed on it didn't catch it on fire. They would invite the children from a local orphanage to sleep in the spare bedrooms of the mansion, while spending a few days at the resort swimming in the pools, and playing about the grounds as a Christmas treat. In 1940, Dutch's maternal grandmother came from Malang to live with the family, as she had become an elderly widow. "Oma", as the kids knew her, was the loving, nurturing maternal figure that her daughter Carolina most surely was not. ("Oma" means "Grandma" in Bahasa.) The children remember her as a beautiful older woman, with a loving, gentle and protective spirit, always ready with a hug, a kind word, and dispensing out gentle correction when needed. She was much beloved by them all.

As Jan grew, he developed an especially close bond with his sister Patricia, who was nearest to him in age, but six years older. Patricia doted on Jan, and took up much time with him, teaching him basic concepts, words (in the five languages,) and skills. She helped with his daily care and feeding, and obviously loved him dearly. As he was yet too young to attend school, she enjoyed teaching him what he could learn at each stage of his development, and with his innate intelligence, he was a fast learner. For some reason, Carolina seems to have been unduly harsh in her dealings with Patricia. Carolina was prone to fits of moody rage (which sometimes occasioned her throwing and smashing household items, beating at her husband with an umbrella as he sat placidly, just shooing her off, and other abusive behaviors.) During some of these depressive episodes, she would fly upon Patricia with harsh criticism and severe punishments for trivial or imaginary offenses. A common punishment was to lock her below stairs in the servant's quarters, in effect imprisoning her for periods of time. Even then, Patricia had a strong artistic bent. Her method of coping with this isolation was to paint beautiful murals on the walls to pass the time. As her mother rarely if ever ventured into the servants' quarters (which she no doubt saw as being beneath her both in status as well as actual elevation) it appears that Patricia did this without suffering additional consequences for it.

Emil showed an unusually broad-minded sense of racial tolerance for a man of his time and parentage. *"My dad was one of the first Dutch settlers that let his kids play soccer with the servant's kids,"* remembers Dutch. When he was still a young boy, his sister Patricia went looking for him one day. She looked across a field, and saw a whole crowd of little brown-skinned kids crowded around a water buffalo, and sitting on the buffalo's back, one little blonde kid. Of course, it was Dutch, placidly unafraid and having the time of his life. *"It was a fun life!"* is how Dutch describes it. Always given some pocket money to sport around, young Dutch would impulsively reach into his pocket and purchase any toy or curiosity which caught his privileged eye during the day in the hands of a servant or other child. In the evening, when Oma discovered his new booty, she would take him by the hand and walk him around, making him return the items to their original owners. Young Dutch grew steadily, and began to demonstrate the traits of intelligence, bravery, strength and patience which in time he would come to exemplify. He was an engaged child, learning rapidly and developing into an inquisitive and confident boy. He faced a potential future of moneyed ease, advanced education and a privileged lifestyle. Then, in May of 1942, everything abruptly came crashing down. Dutch was not yet six years old, as his birthday was not until the end of July.

CHAPTER 5:

RISING SUN

In the early years of the 20th Century, Japan had begun laying the groundwork for their eventual planned domination of East Asia. Starting by expanding business and trade links, Japan ultimately proposed the establishment of the "Greater East Asian Co-Prosperity Sphere", which they initially portrayed as a type of free-trade zone under Japanese leadership. In time, their naked ambition was made clear, and they embarked on a campaign of outright invasion and domination of several East Asian nations. Stoking the fires of resentment against foreign colonial powers, Japan promised "Asia for the Asians" and encouraged the smoldering nationalist sentiments which permeated several East Asian societies, including Indonesia. (They particularly encouraged Indonesian nationalism and anti-European hatred in Java, planting the seeds of a greater holocaust which was later to be manifested.) On the 8th of December, 1941, the Dutch Government-in-Exile (which was already at war with Nazi Germany) declared war on the Empire of Japan. This was in response to Japan's expansion of the war through the attack on the US Naval Base at Pearl Harbor, Hawaii the previous day, which also resulted in the US's immediate entry into World War II. Japan's response was ruthlessly swift, and in January of 1942, they began invading parts of the Dutch East Indies. By the 1st of March, they had landed unopposed in four locations on the coast of Java. And by the 9th of March, the Dutch Colonial Government had surrendered.

The Japanese occupation was initially greeted positively by many Indonesians, who met the invading troops waving flags and voicing their support by shouting slogans such as "Japan is our elder brother!" Many Indonesian malcontents took advantage of the Japanese forces' advance to murder Europeans, especially Dutch citizens, who they viewed as symbols of the Dutch Colonial Administration. They also informed the Japanese as to the whereabouts of larger groups of European citizens. In contrast, some other Indonesians showed touching loyalty to their Dutch neighbors, helping to protect and warn them.

As was their pattern across their captured territories, the Japanese immediately set about establishing internment camps for the imprisonment of those who they perceived as enemies. Eventually, they interned over 100,000 European and Chinese civilians, and 80,000 Dutch, British, American and Allied troops in Prisoner of War camps. In these camps, death rates averaged between 13 and 30 percent of internees. Discipline was harsh and physical sustenance was minimal. In areas which were heavily occupied by Japanese troops due to their being perceived as strategically important, arbitrary arrest and execution, physical torture, sex slavery and other war crimes were widespread. Many millions of Indonesians and European settlers were also kidnapped and exported from Indonesia as forced laborers for Japanese military projects. Estimates range from four to ten million people in Java as having been forced to work for the Japanese military. A post-war United Nations report stated that four million people died in Indonesia as a result of forced labor and famine, including 30,000 European civilian internees.

CHAPTER 6:

MIEL'S RELOCATION

In May of 1942, things were getting tense in the Versteegh family's home. The Japanese Army had taken the capital of Jakarta, the Dutch Colonial Government had surrendered, and the Japanese had begun their Administration of Occupation. The Japanese Secret Police (the dreaded *Kempeitai*) actively collected information on European settlers of note. Before the area had been completely subdued, Emil Versteegh and his eldest son Joop had enlisted in the local territorial units of the Dutch Army, which they perceived as a patriotic duty. The Army assigned these military forces to combat the advancing Japanese. They were not posted to the area near Kalibening, and the family did not know their exact whereabouts. This left 15-year-old Miel as the oldest male of the family at home, with Carolina and the younger children trying to carry on life as usual under the shadow of Japanese occupation. Given that the resort was a fairly strategic property, with nice facilities directly on the Kalibening River, it attracted Japanese interest. On a few occasions in the Spring of 1942, Japanese troops visited the property. When they came, they encountered young Miel. A strapping, extremely powerful teenager, Miel enjoyed wrestling, and somehow got involved in informal wrestling bouts with some of the Japanese soldiers....who he soundly defeated in these contests of strength. Ultimately reaching 6'3" in adult height, Miel even at that time enjoyed both strength and leverage advantages over the more diminutive Japanese. He actually threw a few soldiers into the swimming pools while wrestling them, to the jeers of the others. The Japanese soldiers were not amused by this loss of face, and soon realized that Miel was not just a kid; he was a formidable potential adversary, as well as a useful engine of forced labor. They arrested Miel and took him away under guard. Soon thereafter, Miel was herded along with over a thousand other prisoners down to the docks. They were placed into pig crates and loaded onto a cargo ship bound for forced labor camps in Japan. Miel was fortunate; he was one of the last loaded onto the ship, so he was higher in the hold and closer to air, sunshine and rainwater. This is why he survived. Most of the other prisoners who were loaded lower in the ship died of starvation, dehydration or disease. The Japanese provided them no care whatsoever on the ocean journey to the Japanese homeland. Although Miel survived the war, he never returned to Indonesia and Dutch never saw his brother again after his arrest in 1942.

The realities of the Japanese occupation and its concomitant severity had begun to be felt among the Indonesian populace, and in many areas, guerrilla resistance cells formed and began to harry the Japanese forces. Partisan groups arose in the hills above Kalibening and began conducting sniper attacks, ambushes, and acts of sabotage against the Japanese. Carolina was of course supportive of these actions; her husband and stepson were Dutch soldiers and her other stepson had been conscripted and taken away by Japanese troops. When she was approached by a neighbor asking for her help in support of the guerrillas, she agreed to provide them what she could. She began cutting bed sheets from the resort's hotel rooms into roller bandages, which she passed to the resistance with some other foodstuffs and supplies. In this way, she became an active participant in the conflict. Due to the divided loyalties among the populace, some pro-Japanese native no doubt informed on her. At any rate, she came to the attention of the Japanese administration, who ordered her internment with the remainder of the family. But divided loyalties ran both ways;

sympathetic neighbors informed her that the Japanese had been asking questions about her activities, and they rightly intuited that it would not be long before the family was arrested. So they worked out an early warning system.

CHAPTER 7:

RUN TO THE HILLS

The Versteegh resort was many miles outside the city of Magelang, up a four-mile road from the last intersection with a larger highway heading to the city. The crossroad was home to a small village of local houses. Soon enough, a detachment of Japanese soldiers in an Army truck turned at the intersection and headed towards the resort. When this was observed, a resident moved outside and began a discrete pattern of rhythmic drumming on a traditional Indonesian drum. This sound carried up the valley quite clearly, and was heard at another home some distance up the road. That occupant took up the percussive signal, and in this way, an audible warning of the approach of the Japanese was heard at the Versteegh home, moving at the speed of sound and arriving well in advance of the soldiers. When the sound of drums was heard, the family sprang into action. Spreading a bed sheet for each member of the group, spare clothes, food, water and necessities were quickly tied into a bundle and thrown over one shoulder. And Carolina, Onno, Millie, Patricia and five-year-old Jan quickly walked into the hills above the resort, accompanied by a few servants to see to their care and protection. By the time the Japanese arrived, they were out of sight and getting further away every minute. When queried by the Japanese detachment commander as to the family's whereabouts, another servant only replied that the family was away, and they did not know when they planned to return. The property was searched, nothing found, and in time the soldiers departed and the family was sent word that it was safe to return.

Dutch remembers the frantic walks into the hills to escape the Japanese quite clearly, along with the sense of fear which they engendered. This pattern worked for a while, and was repeated three or four times before the Japanese figured out that they were being had. So on the next unproductive visit, the Japanese commander assembled the servants and informed them that he would return again soon. If the family did not give themselves up, some servant would be executed on the spot, and the process would be repeated until the family was found. When Carolina returned to the home after this visit and was informed of this, she knew that she could not allow the murder of one of the family's servants in an attempt to avoid her own arrest. And within a few days, the soldiers returned. Carolina bravely surrendered herself and the children to the Japanese. They were loaded onto the Army truck under guard, with only the clothes on their backs as personal possessions. And so began a two-year imprisonment.

CHAPTER 8:

THE JAPANESE INTERNMENT CAMP. 1942-1945

The Versteegh family was trucked for several hours over increasingly remote roads to a large internment camp in the hills near Muntilan, outside the larger city of Magelang, Java. This was one of Japan's major internment camps in Indonesia during the war. Onno, as an adolescent male, was taken to a separate camp for men and boys, while Carolina, Millie, Patricia and young Jan were placed in a camp for women and small children. It is difficult for 21st Century Americans to imagine the squalor and deprivation which this camp embodied. You could not escape it, as it was all surrounded by barbed wire and armed guards. Temperature and humidity were utterly uncontrolled. When it was hot, it was sweltering. When it was cold, it was frigid. Rain was an opportunity to catch fresh water for bathing, drinking and washing. Mud, insects, and disease were all around, privacy and personal possessions were minimal. And violence hid around every corner, as people under the extreme stress of captivity reacted in predictably savage ways.

DUTCH REMEMBERS: *"The camp was a former school with brick buildings, now surrounded by barbed wire and guarded by Japanese soldiers. They patrolled outside the wire mostly, armed with Arisaka rifles and bayonets. We were in that prison camp for 2 years. It was a real nasty time. We slept on the cold concrete floor in a corner of a room, on straw mats or whatever blanket we might have. We had little food, mostly rice and thin fish stew. People outside tried to send in food for others, if they could. Oma, as a native Indonesian, was not arrested, and she came to the camp to bring us food when she could, passing it through the wire to us. She was old and not in good health, but she came and did what she could for us.*

We kids had to clean the latrines, pick up garbage. You had to clean the latrines barefoot, because even if you had shoes, some other kids didn't, so they made us all work barefoot. It was a dirty environment, with considerable hunger, disease and mistreatment seen on a daily basis. There was a demarcation line between north and south in the camp marking separate territories and we fought each other. It was just a big melee, and the Japanese didn't do anything about that. That was prison camp life. I once almost killed a kid. Not long after we got to the camp, he tried to steal my shoes. He didn't have shoes and I did. I was choking him, they had to pull me off of him.

(Author's note: It is important to remember that at this point, Dutch was SIX years old.)

If you were caught stealing, or otherwise breaking the rules, they would put you in a steel cage for some period of time, exposed to the elements. Conditions in the camp were so bad and got so much worse over time that people got desperate. Some of the people would try and smuggle the kids out in garbage bags, crates of garbage like pig crates. It went right the first couple of times, but the Japanese got wise to it, and started sticking their swords and bayonets into the garbage. They got a couple of kids, so we stopped. Once, early in my time in the camp, I had been walking behind a Japanese guard holding a stick, and playing like I was going to poke him in the buttocks with it. I moved the stick at him right as he came to a halt and did in fact poke him. That did not go well for me. It was seen as extremely disrespectful. I spent three whole days in the cage once, but not for that. I got punished in the cage three or four times for misbehavior, I was a repeat offender! I was never beaten or kicked or otherwise abused by the Japanese beyond that. Others were, but not me."

The family remained in Muntilan Internment Camp until 1945. According to the strategy pursued by the Allied forces (and over General of the Armies Douglas MacArthur's objections), no invasion of Indonesia was ever mounted. Indonesia was simply bypassed and ultimately, in 1945, Japan surrendered. The Japanese released the Muntilan internees, including the Versteegh family. Carolina led the children out of the gate and they made their way to the nearby city of Magelang. They had the tattered clothes on their backs and few other resources. Carolina found the house of a Dutch friend, now unoccupied, as the family had fled Indonesia before the Japanese arrived. She assumed that the friend would not mind, and moved the family into the home as they set about trying to locate Emil, Joop and other members of Emil's family, while obtaining food, clothing and other needs. Unfortunately, the family learned that Carolina's mother Oma was hospitalized and in ill health. As Patricia was in good physical shape and bore a more native appearance than the other women, she was able to make her way safely to the hospital and see Oma once more before she further deteriorated and ultimately died. The other family members never saw Oma again after the day of their initial arrest by the Japanese, other than her few visits to the camp to bring them food. And in the collapse of Japanese Occupation Government, the fires of nationalism and resentment against European settlers which the Japanese had so cynically stoked were seething, and about to explode into a bloody, five-year civil war.

AS DUTCH RECALLS: *"When the Japanese surrendered, the Indonesians didn't know what to do at first, so they let us out. We had no money, no real contact with the outside world. We lived for 2-3 months in a house of a friend who had gotten out of Indonesia before the war, so their house in the city was vacant. When we arrived, there was a dead body on the porch; we had to remove that. The Indonesians were already planning to kill as many European settlers as they could, in their pursuit of full independence. We were invited, but did not go on a "free trip out west", which they billed as a "rescue convoy" that had been organized to evacuate some people. That was lucky; the convoy was ambushed by the Indonesian forces and they were all slaughtered."*

Unbeknownst to Carolina and the children, there was some very good news. Her husband Emil and her stepson Joop had also both survived the war, against very long odds. Both Emil and Joop had been imprisoned by the Japanese authorities, having been captured when Dutch Army elements in the Far East had surrendered. They were separated upon their capture, and neither had any idea where the other was being held. Emil was first confined in Singapore's infamous Changi Prison, the scene of the book *"King Rat"*, by James Clavell. (Clavell was himself an Australian POW in Changi Prison, so wrote the book from first-hand experience. It was later made into a movie of the same title in 1965.) Conditions in Changi were harsh, but generally survivable. But Emil would not remain there. When he was found to have been a part of a group of prisoners who had been building and operating clandestine radios and planning escapes, he was transferred to one of the absolute worst scenes of Japanese atrocity in the entire history of World War II. Emil was sent to a forced labor camp at Katchanaburi, Thailand, where among other construction projects, a certain bridge was being built – the famous Bridge on the River Kwai. (The corresponding eponymous 1957 motion picture, starring Sir Alec Guiness, depicts some of the horrific aspects of the labor camp at Katchanaburi, but vastly understates them due to cinematic limitations.)

Japan needed to build a railway to link their new imperial possessions in Thailand and Burma for strategic and logistic purposes. This would require the construction of a one meter-gauge railway from Ban Pong, Thailand to Thanbyuzayat, Burma. The line passed across 250 miles of extremely difficult jungle and mountain terrain. It is now known as "the Death Railway." The construction was accomplished by several thousand Allied POWs and Asian slave laborers, who were held in extremely primitive and harsh conditions. They were forced to live in open bamboo thatch shelters, given very little food and water and no medical attention. They were basically worked to death in extremely dangerous conditions and subjected to especially (but for this war, typically) harsh disciplinary measures by their Japanese captors. The work began in October, 1942 and was completed in one year. It is estimated that for every single crosstie laid on this 250 miles of track, one laborer died, totaling over 9000 men. A War Cemetery at Katchanaburi, interring approximately 7000 of the laborers, and a second at nearby Chungkai where another 2000 lie, give mute testimony to the unspeakable cruelty enacted here. A small Death Railway museum is maintained at the site of the famous bridge today, documenting these crimes against humanity and memorializing the hardships endured here by the prisoners. The author has visited the site and seen it with his own eyes.

In this miasma of maltreatment, Emil (who in 1943 was a forty-two-year-old man) began to founder. The combination of extreme labor, exposure to jungle conditions, poor rations, zero health care and general despair all combined to reduce him to a state of near death. He was days away from joining thousands of other predecessors in a shallow, unmarked grave. And then a miracle happened. Emil's eldest son Joop, who had upon his capture been been immediately transferred to the Burma Railway project at Katchanaburi, happened upon his father lying on his near deathbed.

DUTCH SAYS: *"I always made fun out of it, but it's true... Joop had healing powers in his hands."* Joop (who would eventually go on to become a trained Medical Nurse) gave his father part of his limited rations, worked to improve his deteriorated physical condition, and nursed him back to health using his natural powers of healing. And both men survived to be liberated and ultimately repatriated to Indonesia after the 1945 surrender of the Japanese.

Emil's second son Miel had likewise survived the war. After surviving his harrowing shipment to Japan in a pig crate aboard a freighter, he had been assigned to a labor battalion and put to work on several war-related tasks while under Japanese imprisonment. As they were supporting the war effort, these men were given survivable conditions and sufficient food to keep them viable as laborers, and Miel had in any case always been a particularly hardy and robust physical specimen. He lived on and would ultimately emigrate to Australia.

Carolina's eldest son Onno had also survived the camp for men and older boys where he had been imprisoned in Eastern Java. Given the percentages of fatality in the Japanese internment camps, the fact that the entire family survived may be seen as nothing short of miraculous. But Onno's whereabouts were unknown to the family at this time.

For the moment, Carolina was unaware of any of these things. Despite the divine blessing of survival and after only a brief period of freedom and recovery in the aftermath of Japanese internment, the Versteegh family's fortunes again nosedived, in an even more horrific fashion.

CHAPTER 9:

HELL WITHIN BARBED WIRE. INDONESIAN INTERNMENT, 1945-1946

Within about three months from their release from Muntilan Camp, the Versteeghs were again arrested by forces of the new regional Indonesian administration. By this time, a full-fledged program of intimidation, robbery, kidnap, targeted murder and organized massacres was underway. This dark period of Indonesian history is called the Bersiap, and it raged most severely from 1945-46. The death toll estimates range in the tens of thousands of innocents, mostly European and Chinese foreigners who were declared "Musuh (enemies) of the people" and killed by beheading, mutilation and sexual tortures and other savage means, often in grisly public spectacles. Victims' bodies were usually disposed of by throwing them in the sea or nearby rivers, and other than some mass graves holding a few thousand bodies, most of the victims have never been found. More than twenty thousand registered Indo-European civilians were abducted, and never seen again.

When they were again arrested, Carolina, Millie, Patricia and Jan were taken to an internment camp in Magelang, run by Indonesian independence advocates in positions of administrative power. They entered the gates with the clothes on their backs. The gates were locked behind them, and the darkest period of Dutch's life began. The facilities were rudimentary in the extreme, and the Indonesians provided little to nothing in the way of supplies or rations. It was, quite simply, a holding pen for the internees to die in.

AS DUTCH REMEMBERS: *"The Indonesian government forces came and arrested us and put us back into a concentration camp for two years, because we were Dutch foreigners. They hated foreigners, and wanted to kill us off. I was eight years old when we went in. The camp was on the site of a former military unit, a cavalry unit. It was really ramshackle, surrounded by barbed wire. We were housed in the old wood frame stables on dirt floors. There were big gaps in the boards, no screens on the windows, no beds. It was far worse than the Japanese camp. Whatever you were wearing the day you arrived is what you had to wear. We were filthy, and ragged. They gave us practically no food. We were eating paste made from laundry starch, and whatever else we could find. My sisters and I tended a little patch of grass, and we would go gather grass and cook that up to eat. I did learn not to like escargot; we had to eat a lot of it. I don't know why people like it. I was suffering from acute malnutrition and beriberi. I had so much fluid in my body, it would collect on one side when I lay down, and when I turned over, it would drain down to the other side. I was dying. An old lady had some vitamin pills and gave them to my mother. She said "I'm old, I don't need them – he does." She saved my life, really. The Indonesians would severely beat people with sticks, usually out of sight of everyone else. This happened a lot. Some of them died. Lots of other people just died of starvation, or sickness, or just despair. The Indonesians would also take anyone who arrived at the camp or were in the camp* *and if they thought they had money or valuables, they would just publicly behead them. I saw the bodies and the heads near the gate frequently. There were dead bodies everywhere, often."*

Let that sink in. The above is the eyewitness account of a boy of eight to ten year's age at the time. Imagine the mud, the flies, the stench. Visualize the sights of savagery, brutality, death, and decay. And consider the impact these realities must have had on the spirit and consciousness of anyone subjected to them. Through all this, Dutch's sisters and mother did the best they could to keep each other healthy and alive. The girls, Patricia in particular, did what they could to continue providing a rudimentary education of informal lessons to young Dutch as they all struggled to survive.

In the immediate aftermath of World War II, most of the world's countries knew little and cared even less about the humanitarian crisis of a sprawling Southeast Asian country. The Dutch Government had no ability to effectively govern the post-war territory which was aflame with nationalist and xenophobic sentiments. So a cruel civil war raged while most of the rest of the world averted its eyes. There were few efforts made to halt the widespread atrocities. In most Western countries, "the boys were home from the war" and there was little to no appetite for involvement in another foreign adventure. But the situation grew so unstable and the killings and conditions so horrific that in time international efforts to establish order simply could not be avoided. Two major diplomatic interventions were conducted, and Dutch and former Allied armed forces eventually began to re-exert control, at least in the major cities. They never succeeded in enforcing order and calling a halt to the killings in the countryside. By 1949, the situation seemed to be irreparable, and international pressure on the Government of the Netherlands forced their abandonment of efforts to retain possession of their former East Indies colony. The Netherlands recognized the independence of Indonesia that year.

Dutch remembers their release from this hell. *"We were in this camp for about two years, until we were liberated by British armed forces. We were discharged from the camp. We were in very poor shape."*

CHAPTER 10:

GURKHA RESCUE

Upon their release from the camp in late 1946, Carolina and the children were in awful physical condition, isolated and with few resources. They limped into the nearby town of Magelang. As they had during their previous brief period of freedom two years earlier, Carolina found the vacant home of a Dutch friend who had escaped Indonesia before the Japanese occupation, and installed the family in it while she tried to find out the whereabouts of her husband, stepson and other relatives. Despite international efforts to reestablish control, a state of tremendous prejudice and violence against foreigners still existed in the town and surrounding countryside. The family were the hated Dutch enemy, and their lives were at extreme risk. With their fair skin and light eyes, it was not safe for Carolina or Millie to leave the house and go about the streets in Magelang. But Patricia and young Jan could and did go out in search of food and information, with their more native-looking skin and eyes providing them natural cover. Patricia's bravery and success in these foraging expeditions likely saved the family's lives, allowing them to hold out in hiding until their eventual rescue.

In their efforts to establish local control, The British had stationed in Magelang a detachment of one of their most famous regiments – the Gurkhas. These ethnic Nepalese soldiers were and are justly famous for their extreme tenacity and ferocity in battle. They have carried the day in countless battles on behalf of the British Empire, and their regimental records are replete with numerous awards of the highest decorations for bravery and victory. The Gurkhas are quite justifiably counted among the world's most renowned and feared military units. Their regimental fame continues to this day. Gurkhas traditionally carry a unique edged weapon as part of their battle kit. Called a kukri, it is a curved, forward-weighted leaf-shaped blade with a razor-sharp edge and a flared buffalo horn handle. It is a formidable chopping and slashing implement, capable of hacking off an arm or head with one stroke, and the Gurkhas take tremendous cultural pride in maintaining a high degree of skill with it.

The political situation was extremely unstable, and the Indonesian rebels set up armed checkpoints on several roads, asserting control in defined areas outside of the sectors controlled by the British. One of these checkpoints was set up at an intersection near the house where the family was staying. Carolina still saw herself as a loyal Dutch citizen, and quickly ran afoul of the local Indonesian independence guerrillas, informing on their checkpoint activity to the temporary British security administration. This attracted the guerrillas' attention, and almost cost the family their lives in retribution.

DUTCH PICKS UP THE NARRATIVE: *"After release from the Indonesian camp in 1946, we were free and managed to find a Dutch-owned house on a corner, outside of the town of Magelang. The owners, who my mother knew, had fled Indonesia before all the trouble started. We lived there for several weeks, in great danger. I was about ten years old. There was an Indonesian checkpoint right down the road from us. My mother used to tell the occupying British regional security forces what they were doing. They found out about it, and decided to do us in. My* mother had us try to just hide in the house. We barricaded the doors and windows with furniture, pianos, whatever we could find. The Indonesian forces would just shoot straight into the house, bullets just passing through everywhere. The house was completely riddled. We lived in the iron bathtub a lot, seeking ballistic protection. They would come, beat on the doors, and yell at us. "We know you are in there!" But they never came in, or burned the house with us in it. Why, I'll never know.

One day, my mother heard voices in the back yard. She heard English people speaking English! She took a chance. She opened the back door, and it was a detachment of Gurkha soldiers, under the leadership of a British officer. She told them what had happened. They took us out of the house, and that's when I learned how to low crawl...underneath a Gurkha! People suddenly were shooting at us, at the Gurkhas, everybody. We were crawling under intense fire. That's when I learned how the Gurkha soldiers kill people – with their famous edged weapon, the Kukri. An Indonesian insurgent charged right at us, trying to kill us. The Gurkha had a kukri on a lanyard, and had to just swing it around his head quickly and release it out, letting it extend to the end of the lanyard. It just cut the man's throat. It happened so darn quick. They fought us clear of the house under fire. They led us from house to house evacuating everybody that was in danger.

They led us to the airfield, and we were flown to Surabaya. They flew us on burned, shot-up leftover World War II cargo airplanes, it was unbelievable. We had to take showers, have delousing powder put on us. We went to Surabaya, and then Jakarta. My two brothers (Joop and Onno) and my father were already there and that's where we were reunited as a family.

CHAPTER 11:

REBUILDING A LIFE. 1947-1952

The family was overjoyed to be brought back together again after five years of separation and unspeakable hardship. They slowly regained their health and began to try and resume something like a normal life, amid the smoldering racial and national tensions attendant to the ongoing Indonesian civil war. The family's economic circumstances had been vastly damaged, and the days of moneyed privilege were over. The marriage between Emil and Carolina was fraying badly. Emil was given a minor government job as a regional zoning officer for a time. Unfortunately, his assigned work partner was corrupt. When fiscal irregularities were discovered, Emil was suspected or jointly blamed, and he lost this position. So when a nephew of Emil's named Ignaz Schmutzer offered to let Emil help manage another extended family-owned rubber plantation in Western Java near the city of Bandung, Emil had little choice but to accept and relocate himself there. Carolina and the children took up full-time residence in Jakarta, and occasionally visited the plantation to see Emil and escape the capital city. This separation was not simply driven by economics. Emil and Carolina had also discovered that the years of separation and stress had taken their toll on the marriage. As they were married in the Catholic tradition, divorce was not an option, but they were no longer married in spirit and never considered themselves to be so again.

Ignaz Schmutzer was an interesting character. He was tall and wiry, with dark hair and a closely trimmed beard and mustache. Independently wealthy, he lived a life of adventure worthy of its own tale. He had flashy cars, clothes, money. He had been married to a Spanish royal princess. During World War II, he had worked with the anti-Nazi underground, hiking through the mountains of Europe trying to rescue and exfiltrate Allied airmen who had been shot down, returning them to friendly territory. He was possessed of a good deal of flair and `elan. Dutch says *"He was a ladies' man. Oh, hell yeah. He really knew how to get around with women."*

Bandung was an important city which had a long history of Dutch habitation, having first been colonized in the 18th century as a center of tea cultivation. It lies 2520 feet above sea level, approximately 87 miles southeast of the modern-day capital of Indonesia, the city of Jakarta. It is situated in a river basin, and surrounded by volcanic mountains which afford the city excellent natural defenses, so much so that the Dutch East Indies government planned to move the capital from Jakarta (then known as Batavia) to Bandung. It also features cooler year-round temperatures than other major Indonesian cities. By the early-mid Twentieth Century, Bandung had developed into a resort city for wealthy plantation owners and visitors. Its cafes, restaurants, European boutiques, and luxury hotels had earned it the nickname "Parijs van Java" (The Paris of Java.) But it was much diminished in the aftermath of three years of Japanese occupation and subsequent civil insurrection against Colonial Dutch authority.

Dutch now found himself entering a new and strange phase of life: that of a normal adolescent boy who was well behind in his formal schooling! His sisters and other prisoners had taught the children the fundamentals of reading, writing and arithmetic in the camp, so he rejoined his age group in school about age 11, circa 1947. But Dutch was suffering the quite predictable symptoms of the extreme duress of his years of captivity, undoubtedly afflicted by what we now know as Post Traumatic Stress. His sisters remember that young Dutch hardly spoke at all for a period of about three years after the family's release. He communicated in the most minimal way possible, volunteering no conversation and answering only when absolutely necessary. It took time for these invisible wounds to heal sufficiently for Dutch to converse comfortably. *"I didn't have anything to say."* is Dutch's commentary on the matter.

DUTCH REMEMBERS: *"After that, we stayed pretty loose. We didn't have the resort at Kalibening anymore, it was confiscated by the Indonesian Government. The walls were about so thick, concrete. It had a pretty big house. It was like a mansion, really. It had thirteen bedrooms, ten or eleven bathrooms. During the war, it was flattened to the ground. Because we had money, the local people thought we had hidden it in the walls. There was no money in the walls, but it was all destroyed; they knocked all the walls down looking for it. The Indonesian Army took the property over. It is now their Officer Candidates' School.*

I started school in the fifth grade, after the war. I never went to grades 1-4. My dad said "you will graduate in the normal time" and you did whatever he said!"

But Dutch's re-immersion in formal education was not without a hiccup or two. He did not care for school, and had developed a bit of a wild, rebellious streak somewhere along the way during four years of prison camp life. His account of his re-acclimation to formal learning demonstrates the difficulty he had in adjusting to routine life for a war-hardened boy of now eleven years:

"My dad changed my life, because he sent me to private (Catholic) school. I got kicked out of public school. We were a bunch of war kids! After growing up in a concentration camp, I wasn't very interested in studies. We had walls of open bamboo thatch, and

when classes got too boring, we just carved a hole in the wall, went out and went to the railroad tracks. Before school was out, we'd pop back up. And I ultimately got kicked out of the school. So, my dad took me to the St. Ignatius Catholic school in Jakarta. And the priests took care of business. My uncle (a Jesuit priest) was the Dean of the school. He had come to Indonesia after the war. My aunt, who was a nun, was the Dean of the girl's school. We didn't have any choice. My sisters went to the girls' school. After a time, I was playing like eleven different sports. The Dean of the school (my uncle) called me in, and said "well, I see your grades are pretty much down"...what could I say? I'd rather be on the sports field, or somewhere else. So he said "what are your best sports? You can keep three of them." So I chose swimming, soccer and fencing. (My dad was a fencing instructor in college.) If it wasn't for that school, I would never be here. They made a man out of you. They either kicked it, whipped it, or whatever else into you. Study was not done at home, it was done at school after hours. At five o'clock, you came back to school and studied under supervision until about eight o'clock pm. It was a good thing, it really was. So, I graduated! In those days, if you had an A/B+ average, you didn't have to take the final exam. I got it up to that point...and besides that, I had a broken right arm from sports- playing rugby, I think. I was kayaking with my buddy when everybody else was taking the test!"

Religious instruction was also part of the learning Dutch was expected to assimilate under his Jesuit uncle's strict tutelage. He humorously remembers: *"I used to be an altar boy!"* with a snicker. His uncle had insisted that he needed to fulfill this role. It was not a good fit, to say the least. (Dutch is no longer a practicing Catholic, having become disillusioned with the Church's historical missteps of behavior and policy.)

Dutch graduated from the Jesuit-run St. Ignatius School in 1952, at age 16. He had never attended grades 1-4, and caught up to his peer group only through hard work and natural aptitude (albeit with a good deal of enforced discipline.)

Graduation notwithstanding, Dutch's life education continued, even as the family's structure changed a bit. His elder sisters Millie and Patricia both came of employment age and took jobs as Flight Attendants for KLM, the national Dutch Airline. They worked aboard in-country flights in Indonesia. Millie ultimately met and married a KLM pilot, and emigrated to Holland to be with him. Patricia likewise met and married a businessman who ran a soy plantation, which produced the tofu and textured vegetable protein which are so much a part of vegetarian and Asian cuisines. She eventually emigrated to the United States. Onno had likewise come of age and moved on. But now-teenaged Dutch was still at home with his mother, living in Jakarta, or occasionally visiting his father who was living full time on the rubber plantation outside Bandung. And he was always looking for action, and finding ways to get into trouble. His relationship with his father seems to have been a distant one, characterized mainly by intermittent episodes of corporal punishment administered by Emil with, Dutch shrugs and admits, just cause:

"My whole life, I didn't really have that much connection with him. Except when he told me I had to finish school on time! My mother would tell him to come home and discipline me, she couldn't control me! He would come to Jakarta for a visit, and he would whack me a couple of times. It didn't work. One time, I was doing handstands on a motorcycle, and cursed out the driver of a car that tried to cross me....it was my dad. That didn't go over well."

Dutch was not the only of Emil's sons who he had difficulty controlling. As Dutch's older brother Onno was now of age, Emil arranged to send him to Holland to complete a formal college education. But Onno had other ideas. Instead of attending classes, Onno spent the tuition and subsistence money Emil provided him on an extended multinational period of carousing and debauchery. Partying and skiing in Denmark were much more interesting than Catholic college. *"That was my college money he spent!"* is Dutch's wry commentary on the matter. When he ran out of money, Onno was inducted into the Dutch Army as a young private....and sent back to Indonesia!

CHAPTER 12:

FIRST COMBAT PATROLS. 1952

Dutch remembers this post-war period of about five years as being extremely formative, partly because it took him the first steps down the road leading to his ultimate profession as a warrior. During this time, the newly-independent government of Indonesia was still taking advantage of every opportunity to supplant their former Dutch colonial masters, and to do so, they casually exercised the innate corruption and propensity to violence which characterize most Southeast Asian governments to this day. As the rubber plantations were profitable, given rubber's strategic importance across a wide variety of military and industrial applications, the plantations were prime targets for acquisition or nationalization. It was necessary to drive the foreign owners off, by whatever means could be contrived. In the case of the family plantation outside Bandung, this began as a series of sabotage raids, with an escalating use of destructive violence, which began in about 1952. Growing areas would be burned, rubber sap tapped and stolen, machinery damaged, shots fired, and lives threatened. In an effort to protect their property and interests, Ignaz Schmutser and Emil organized a defense force, arming and training them to patrol and defend their plantation properties against the escalating raids of armed thugs, hired and equipped by senior Indonesian government officials. There was no shortage of surplus weapons in post-war Indonesia, nor of combat-experienced men to wield them. And when, during one of Dutch's periodic visits to the plantation his father asked if he wanted to participate, now 16-year old Dutch was of course excited to volunteer for the defense guard force. He was issued a weapon, taught to fire it accurately and trained in the art of silent jungle patrolling. And he participated in several series of combat patrols during his visits over the next few years.

HE DESCRIBES THE TIME: *"After the war, my dad went back to an uncle's plantation, then being run by his son, Ignaz, a distant cousin. There were four plantations, close together, family properties. These were located above Bandung, West Java, up in the mountains. That's where I learned patrolling. The bad guys would come down and burn this thing down, and then burn this one down, because they were stealing rubber for the President of Indonesia. They were selling it at the time. My dad and my uncle had to run the plantations. So we had security forces. My dad said "do you want to go out with them? When I said yes, he said 'There's your boss, right there.' – pointing to the local man they had hired as the security force chief. I was about sixteen years old at the time. He trained me on how to patrol. We ran combat patrols, sometimes leading other regional security forces. I was issued a weapon – you will never believe. A Thompson submachine gun... which I never touched again after those days. This weapon is.....heavy! You walk around leaning to the side you are carrying it on after two days, it is so heavy. My Dad had told them, 'if he wants to change weapons, he is not going to change. He's going to keep that weapon.' Boy, that was the first and last one I ever carried.* (Author's note: A Thompson SMG weighs 11.5 pounds loaded, with each spare magazine adding about one and a half pounds to the user's load. This becomes burdensome very quickly.) *Yes, I learned my patrolling out there with that guy. We had a few skirmishes here and there with the raiders who were coming to destroy the plantation."*

Imagine 16-year old Dutch, not yet fully grown and naturally slight of build, carrying a .45 caliber Thompson submachine gun while silently patrolling the jungle areas of the plantation with the other members of the Guard Force. He learned the mysteries of quiet, slow advancement, frequent listening stops, careful foot placement, observation using all senses. He gained a strong knowledge of the tactical uses of terrain, vegetation, background noise and ambient light. He gained a firsthand appreciation for camouflage, sound and light discipline, formation and teamwork. This was a contact sport, played in lethal earnest, and the patrols did from time to time culminate in exchanges of fire. This was his earliest experience of a reconnaissance soldier's life, with all its risks. Unfortunately, the greed for the profits of the rubber trade and the deep-seated resentment of foreign owners on the part of Indonesian government officials escalated the conflict of interests beyond the ability of the little guard force to resolve the issue.

CHAPTER 13:

FALSE IMPRISONMENT, AND THE EFFICACY OF BRIBERY. 1953-1954

When it became apparent that Emil and Ignaz would not be easily driven off through the use of force applied by the raids and sabotage, the Indonesian government took things to a whole new level. They arrested Emil and Ignaz on false charges of "supporting insurrection" and threw them into prison in 1953. The properties were seized and ultimately nationalized, with the profits of course being siphoned into the pockets of the corrupt senior officials. Emil and Ignaz spent almost two years in prison, falsely accused, but unable to prove their innocence in the corrupt Indonesian courts. Dutch therefore adapted to the norms of the culture and the time. He engaged the services of a local attorney who had good connections, and converted the once-again-significant family money into a powerful weapon: bribery. Working through the attorney to identify and discreetly grease just the right palms over an extended period, Dutch would meet corrupt court and prison officials in dark and shady bars and literally pass them bankrolls under the table. In this way, he finally secured his father's and Ignaz's release in late 1954. Emil was in very poor shape: as a punishment, he had been tortured by being placed standing in a tiny metal locker for a three-week period, resulting in severe tendon and nerve damage. Emil was in fact partially paralyzed and unable to walk much for about three years after his release from detention. *"He was a hard old guy."* says Dutch.

Unfortunately, the Indonesian Government persisted in its efforts to permanently seize the land and immediately threatened to again imprison the Versteeghs. The situation deteriorated quickly and as they were being actively hunted for rearrest, the family was forced to abandon Indonesia entirely, for good. They went into hiding until Ignaz Schmutser arranged passage on a ship bound for Holland for them, and they headed for the docks to flee the country. But things did not go as planned.

DUTCH RECOUNTS THE TALE: *"Then of course, we lost those plantations, and they put my Dad and my cousin Ignaz in jail. They nationalized the land and imprisoned them on trumped up charges of supporting 'anti-government rebel forces.' This wasn't true. It took me and a lawyer (I was only about sixteen or seventeen years old) to get them out. I walked around with huge rolls of cash money in my pockets...under-the-table money. I was passing large bribes to judges, other officials. I was introduced to that kind of life early. I would visit them in the prison and take food to them, as the Indonesians were giving them very little rations and other necessities. Well, we finally got them out, and ultimately we had to leave in a hurry. So we booked passage and planned to get out on a boat, my mother, my dad and myself. But there was a problem. On the way to the boat, an Indonesian man tried to steal my jacket, or something. I fought him, and put him down. But with my last kick, I hurt myself. I kicked the guy. I kicked him, all right. I kicked him so hard I burst a hernia. I was in agony when they helped me board the ship. The boat captain said "he can't go aboard this ship, he will die without proper medical care." He had no way to operate on me.*

So, my cousin Ignaz took me back. I now had a different surname (Wierenga) than my mother and father (Versteegh) who were being hunted and who then immediately left aboard that ship without me. My cousin took me back to the hospital that had long served my family. The doctor there was supposed to operate on me, but the bastard was always drunk, he couldn't operate. My cousin ultimately found another doctor to operate on me. That was just before Christmas, 1954. I stayed in the country for about six weeks, recuperating from the surgery. My cousin used to bring me food in the hotel, we would eat in the hotel. He told me "do NOT go outside!" I didn't go anywhere. I finally got well enough to travel."

So Emil and Carolina made the sea journey back to Holland and their new home and country without young Jan, arriving in early 1955. Neither would ever return to Indonesia. Their departure from Indonesia did mark one important turning point in family history, though. Before the voyage, Emil and Carolina 's long-simmering marital discord erupted into a full-blown argument. Prior to embarking for Holland, they had to obtain new identity documents, as all their legal paperwork had been lost during the war and subsequent years of captivity. Carolina was given the task of filling out the application forms for the family's new passports. Using this power, she made a unilateral decision, whether out of spite for Emil, or to cement her own family legacy we do not know. But when she filled out the forms for her daughter Millie and her son Jan, she reported their surnames as Wierenga, rather than Versteegh. After only being known as Versteeghs in their early years, both Dutch and Millie carried her family surname for the rest of their lives. Carolina's bureaucratic jab was a permanent slap in the face to Emil, conclusively winning that argument in spectacular fashion.

CHAPTER 14:

WORLD CRUISE! 1955

Dutch recuperated from the hernia repair operation in a hotel room for a month and a half. When he was well enough to travel, his cousin Ignaz was ready to make the arrangements for his final departure. Appreciative of Dutch's many years of service to the family and the plantation, and mindful of the hardships and sacrifices that he had already endured across his young life, his cousin suggested something a bit unorthodox. He had plenty of funds at his disposal, and no doubt with a twinkle in his eye, he approached Dutch with a choice (which was really no choice at all for a brave 18-year-old young man with a taste for excitement!)

DUTCH RECALLS: *"So my cousin Ignaz said, "well, for all you did…do you want to take a trip?" I didn't know what he was talking about. He said, "well, you can go on a passenger liner, and get to Holland in about three to four weeks, or you can go on a cargo ship, and get there in about six months." I thought "well, that sounds pretty good!" So I went on the cargo ship. He gave a slew of money to the Captain, and told him to give me money if I asked for it. The Second Officer was put in charge of me. That ship stopped in every harbor there was along the way, it seemed like. It was fun! He showed me some life!"*

One can almost smell the beer, taste the exotic foods, see the attractive (and readily available) women in the seaside towns and feel the excitement of the bar fights, sea storms and new horizons young Dutch experienced on that memorable extended voyage back to the family's ancestral home. Dutch stepped off the ship in Holland a wise and worldly young man with a whole new future ahead of him as a Dutch citizen recently returned to the motherland. He no doubt had (has) several tales best left untold but firmly cemented in his memory, each of which evokes a quiet smile to this day. Dutch unfortunately lost sight of Ignaz at this point. He knows that Ignaz ultimately lost or perhaps sold the plantation properties, and returned to Holland, but he never heard from him again.

CHAPTER 15:

A SOLDIER, A BUREAUCRAT AND PLANS TO BECOME AN AMERICAN. 1955-1960

When he arrived in Holland, Dutch was not yet 19 years old. He had to do something to occupy and support himself, so in short order, he enlisted in the Dutch Army, in June of 1956. He attended Basic Training at an army base outside The Hague. Given Dutch's long experience with hardship and deprivation, Army Basic Training posed little challenge for him, and his memories of the time are few. Dutch was trained as a field artilleryman, and was domestically assigned to a mortar artillery unit for the duration of his enlistment term. He does remember being comfortable with Dutch Army discipline and lifestyle, but he had bigger aspirations. He completed a two-year enlistment from June of 1956 to August of 1958, and upon his discharge, took a minor government job deep in the Dutch bureaucracy. He lived in The Hague at this period. His vague memory of his duties is shuffling administrative paperwork of some type. But Dutch also had a plan. He had decided that his future lay outside his mother country. He wanted to become an American. So while he worked in his mind-numbing administrative job, he put in an application to emigrate to the United States with the US Embassy in Amsterdam. To help bide his time, he joined a small parachuting club and made a couple of static line parachute jumps, just for fun. It took an additional two years for his application to be processed and approved. But in 1960, Dutch was notified that he had permission to embark on the steamship *SS Rijndam*, bound for New York City and a new life in America. It was one of the happiest days of his life.

"I went into the military in Holland about three or four months after arriving there. I stayed in the Army for two years. In those two years, I put in my paperwork for a visa and permission to emigrate to the United States. It took two more years to get the approvals. But they eventually told me I was ready to go.

At that time, each country could only emigrate 150 people per year to the US. You had to have five hundred dollars in your pocket, you had to have your passport and visa. When I got here, I had to have a job within three months. If I didn't, they would deport me. Five of us came here together, and we kind of had a little gang together. Two of them went back, they just couldn't make it. But I wasn't going to stay in Holland. From the time I was small I wanted to come to America. Why? I don't know. Something just pulled me. I hadn't had any contact with the US Military during the war, except for the pilots that flew us to safety when we were released from the internment camp. But I wanted to be an American."

CHAPTER 16:

WELCOME TO AMERICA! 1960-1963

Like millions of other post-war Europeans, now 23-year-old Dutch took his place in the great American dream. Dutch was one member of a party of five young Dutch men, now fast new friends off on the same life-changing adventure. They were all slated to travel together to their new lives in California. He and they enjoyed a private cabin each, and a smooth eight-day crossing of the Atlantic, and together they stepped off the *Rijndam* in New York on the 14th of March, 1960 and entered the stream of thousands of new arrivals being interviewed, documented and assigned to various cities. It was a sunny and warm day, beautifully welcoming in every way. Dutch remembers seeing the Statue of Liberty for the first time and being quite moved by it. Dutch did not pass through the fabled Ellis Island terminal, as it had been shut down a few years earlier. He showed his passport and visa approvals, was cursorily inspected for general health, communicable disease, and overall suitability to be authorized entry. He easily passed all of these requirements, and displayed the money in his possession, which allowed him the ability to support himself for a time until he could obtain work as was required.

Dutch had intended to travel to California with his friends from the ship, but fate again played a hand. Although intending immigrants were supposed to have shown some basic proficiency in English (which Dutch of course had easily demonstrated, as he had been speaking English in the home since birth), this so-called requirement was observed with a significantly flexible degree of variance. While at the train station with a larger group of new immigrants, he was pulled aside by an Immigration officer. The man conversed with Dutch and noted his solid command of English. Another group of immigrants nearby was in difficulty. This was a group of seven or eight people, comprising three families. They basically spoke not a single word of English...and they were slated to be sent all the way across the USA to California to begin their new lives. So the Immigration Officer made a few notes, changed Dutch's ticketing, and with the stroke of a pen separated him from his new friends. He sent Dutch to California with the non-English speaking group, as the spokesman for them all. This was a very arduous trip from coast to coast. New immigrants were booked in the cheapest accommodations on whatever trains were available, with no thought to comfort and only nodding consideration of basic sanitation needs. The train had no climate control, and only very rudimentary bathroom facilities. Food was very basic. The group sat and slept in turn on hard, splintery wooden benches. And this was when America was still racially segregated. Dutch had never encountered this injustice before and did not understand it when he observed "Jim Crow" policies in action. *"There was kind of a rude awakening for me. I was not aware of the black-white issue at that time. I was in the shower compartment, washing up. There was a black man standing outside, waiting. I told him "come on in!" But he wouldn't. I didn't know about that at the time. We had nothing like that in Holland. That was the first time I had contact with it."*

The group traveled slowly, with frequent stops. And in this spartan and uncomfortable fashion, Dutch traversed the North American mainland, arriving at his new home in California after almost a week of constant discomfort and little, fitful sleep. There, he met his official sponsor, a Presbyterian minister who took him in hand on the train platform and after hosting him for about a week, arranged for inexpensive lodging while Dutch sought gainful employment. The family would invite him to their home many weekends for barbecues, including Dutch into a widening American social circle. (One of their friends was a retired officer of the CIA; this was Dutch's first contact with the Agency, even in concept.) Dutch to this day remains grateful for his sponsor's kindness and assistance. Dutch also continued to pursue his burgeoning interest in parachuting. Building upon his couple of static line parachute jumps in Holland, he sought out and joined an early parachuting club in Calistoga, California, logging many more jumps and expanding his qualifications to include numerous free falls.

DUTCH SAYS: *"I missed Ellis Island by a couple of years, they had closed it two years earlier. They examined our immigration papers, and they were in order. When we got to the train station, the Immigration Officer said, "You speak English, these people going to California don't speak English. You are going with them." And he put me on the train to California, and on the train we went! Sitting in one wagon, on a nasty, hard seat for four days across the US. When we got to California, we were introduced to our assigned sponsors.*

I found my sponsor Norman, and he took me out and we crossed the Golden Gate bridge into Northern California. I was lucky and found a hell of a supporter when I arrived in America. Norman had been a B-25 bomber pilot in the War. He used to tell stories about that. He had then become a preacher, a Presbyterian minister. His wife was from Georgia, and they had no children. She used to tell me all the time 'you have to marry a Georgia Peach!' because she was a Georgia Peach. But I wasn't looking to get married at that time. They were great people, and they lived in Corte Madera, CA. We got together on the weekends quite a bit. I also had another sponsor, a Dutch lawyer for a train company. We got together on the weekends too. They got me into a four-bedroom boarding house on Nob Hill in San Francisco, with my four Dutch friends from the SS RIJNDAM. We partied hard. I lived there for three years.

I had to get a job within three months, or be deported back to Holland. First I got a job as a canvasmaker at a marina, sewing curtains for boats for a short time. All of a sudden, two of the guys had to go back to Holland; they couldn't get a job. So the other three of us continued to live together. Then I worked for Montgomery Wards in Corte Madera, CA., first as a cargo loader, stock boy and truck driver. I reorganized the stocking system, to make it better. The boss came by one day and said "let's go get you a couple of suits" – because they were moving me up into sales. I worked my way up to Assistant Manager of the Men's Clothing Department at Montgomery Wards. They wanted to sponsor me for a full college degree, but I declined, because I thought "if you do that, they own you." I had other plans. I did go to college, a little bitty institution called the College of Marin, while I was working. It was great. I joined a skydiving club in Calistoga, California, and got in quite a few jumps. There were a lot of Army guys in the club. It was all free fall. We used to jump at night, using lines of cars with their headlights on to light up the drop zone. Then we'd quickly get the hell out of there, because we weren't supposed to be doing that."

Dutch's fortunes had definitely improved. He had been taken in hand by a kind, dedicated sponsor, himself a war veteran. He had made some new friends. He had found employment with a major department store chain, and began working his way up the ladder. He invested in his future by pursuing higher education, and was successfully living the American Dream in early 1960's America. But times were changing. The excitement and optimism of post-World War II America had been tempered by another fruitless war against the expansion of the new threat of Global Communism, in frozen and desolate Korea. Then, French Indochina became the new battleground in the struggle between freedom and capitalism, repression and communism. And America began to be concerned with events in another far-off Southeast Asian land, now called Vietnam.

CHAPTER 17:

THE UNITED STATES ARMY. 1962-1966

"After two years, my money ran out. So I said, OK, I owe this country. I'll enlist in the Army. I'll give them four years." Dutch raised his right hand and signed an enlistment contract in September 1962. Officially, Private Jan W. Wierenga attended Basic Training at Fort Ord, California. What this meant in actuality is indicative of a bygone time of Army flexibility. Seeing that Dutch was already a veteran of the Dutch Army, and a qualified free-fall parachutist, the Drill Instructors told him to report to the Ft. Ord skydiving club in Calistoga, and spend his days while packing or trying out several different parachutes!

"For basic training, I didn't do very much. I can say that. They took me out and sent me over to the skydiving club. I'd just sign out in the morning after PT and go jump or pack parachutes. They came out with different parachutes in those days. We were trying out different parachutes. I sometimes brought my parachutes to the barracks and packed them there. Every once in a while I'd look up my Platoon Sergeant. I was in the Airborne-destined group. It wasn't the greatest thing for them to do that – I learned a gentleman's life! It was great!"

Dutch spent almost no time in Basic itself, as he was off jumping and packing parachutes. He attended graduation with his class, but his experience in Basic was quite different from anyone else's! The Army did require Dutch to attend two months of Advanced Infantry Training, also at Ft. Ord. Again, given his previous experience running jungle combat patrols as a teenager and a full enlistment term in the Dutch Army, he breezed through that course and its field exercises. When AIT was complete, Dutch received orders to attend the three-week Airborne School at Ft. Benning, GA, formally qualifying as a US Army paratrooper. Given that this required only five static line jumps, it was no great challenge for Dutch.

"I enjoyed it! It was a good time. The people were really great, there was a lot of camaraderie. The people who were hurt, we helped them out to make it through the Jump School, so that everybody passed. We had a group team leader who was a former fighter pilot, which we didn't know then. He was way older than we were. He had had some kind of incident in a plane, and by the time we graduated, he showed up and had bandages all around his chest. He had been training while injured the whole time."

Dutch made his first five jumps as a US Army paratrooper at Ft. Benning on the 25th, 26th and 27th of March 1963, all from a C-119 aircraft, and the final jump was with combat equipment as was and still is required. The silver wings of a paratrooper were proudly pinned on Dutch's chest upon graduation. Then it was time to be assigned to his first operational unit as a functional soldier.

Dutch had already developed an interest in Special Forces, and he volunteered for this branch. He took and passed all the qualifying tests. *"We took psychological tests, tactics tests, infantry skills proficiency tests, physical fitness tests. I never had any problems with PT tests."* Dutch therefore believed that he was going to be assigned to SF upon receiving his first set of official orders after Airborne School. But when he received his orders, they were to the

82nd Airborne Division, a conventional Airborne unit. When he questioned this, a senior noncom told him "Sorry, Private, this is the Army..The paper says the 82nd. Report to the 82nd!"

"I had jumped several times before that, and they issued me orders to the 82nd . I was a young private, a foreigner, so what do you do?"

Dutch reported to the Second Battalion, 325th Infantry, in the 82nd Airborne. He was there from February through November, 1963. Then he took control of his assignment destiny, but in his own way. *"So, I screwed myself, (not for the first time.) I made my first mistake. I was still mad about not being assigned to SF, so I 'took a short.' I reenlisted a year earlier than my enlistment contract required, and went to Germany. I got married and my wife came about six months later. I couldn't go to war at the time because I was not yet a citizen. We stayed there for over two years.*

Dutch was assigned to the Second Battalion, 509th Parachute Infantry Regiment, 82nd Airborne Division, then based in the "Animal Barracks" at Mainz, Germany. He had signed an early reenlistment contract to obtain this overseas assignment, rather than remaining in CONUS with the 82nd as his initial battalion of assignment went off to war in Vietnam. Upon volunteering, he was assigned to the Reconnaissance Platoon of the 509th.

"I had a great time! Our duties were a joke, really. I started out in the Reconnaissance Platoon. I had made PFC by that time. I had two vehicles I was responsible for – a jeep and a small truck. We were directed to defend the Bulgarian border, in the event of hostilities with the Soviet Union. We expected to be overrun with Russian tanks. I remember we had a big maneuver exercise one time. I was by then a Corporal, with responsibility for five vehicles. We were going through this one town, and the second jeep which had a machine gun loaded with blanks, suddenly opened up with the machine gun. The gunner had spotted a "Russian" car ....and

engaged it with blanks! They probably laughed like hell, it was a big limousine. I had to tell him never to do that again. Later, we were checking out tanks at that time. I learned to be a tanker for about six months, and wanted no more of that. I had hurt my leg and was walking around on crutches while checking out tanks. The Brigade Commander walked by and asked me what had happened, how I was, and looked around. Years later, during the Vietnam War, he was my Division Commander in the 101st Airborne and we saw each other again!

Well, I was single for about six months, until I had the money and bought a plane ticket for my wife and her daughter to come over. I was able to get permission from my Commanding Officer to bring them over, by showing him that I had cash in the amount necessary to pay for both their inbound and return travel to the USA. It was pretty good! We had a little car, we'd take two weekends out of the month to go driving all around Europe. The other two weekends we'd stay home – broke.

When we left, they flew back before me and I traveled back on the USNS GENERAL MAURICE ROSE (AP-126) When I came into Germany, I came in on the ROSE and when I left, I went out on the ROSE. About two or three days out late in the morning, we had this tremendous banging and rattling going on – we thought we had been torpedoed. The ship had lost a propeller. The two days turned into five days, steaming on only one propeller.

Four years came around. It was time to go back to the US. Believe it or not, my wife still has the letter. I got a letter from Montgomery Wards' telling me " we'll hold your job for you for six more months. Make up your mind if you want to come back to work." I said "you've got to be joking. I've got six months left in the Army!" I wrote them a nice letter, explaining that I was in the Army and was about to go to Vietnam. My friends were going to war and I needed to go with them.

CHAPTER 18:

FIRST MARRIAGE

The life of a soldier is a hard one, and wartime makes it even harder. The stresses of Army service and the long separations occasioned by duty assignments put enormous strains on any marriage. Soldiers are not paid well, and it is hard to maintain a family in anything like a comfortable lifestyle on an enlisted man's pay. Many committed professional soldiers, like Dutch, live a "mission first" ethos, which further strains their relationships to and past the breaking point. And many spouses bring their own personal and psychological problems to the marriage. People evolve, personalities grow apart. Divorce is unfortunately common, and over time, marriage takes real work and effort to maintain. Dutch's experience in this regard is very typical of many soldiers'.

"Her name was Catherine (Cathy) Agnes Bailey. We met in a restaurant in Greensboro, North Carolina in 1964. One thing led to another, and I was about to go overseas, so we figured ' we'd best get married.' She just wanted to get away from her father and mother, she was not happy. She had one girl, Dena, who was two years old at that time, from a previous marriage. Her ex-husband used to be mean to them; he was a boxer. It was just an unhappy situation, one of those things. He was, in time, eight

years behind on child support payments. I went to see the judge and I ultimately adopted Dena. He said 'You have two choices. I can make him pay $10,000 (which was a lot of money at that time, but who knew when or how he would be able to pay it), or I can put him in jail for five years.' I said 'I'll take the five years in jail.' They put him away for a while.

We loved each other, we were happy to be together. What with Army life, it wasn't an easy marriage. They came to be with me in Germany, then came home before I did to Ft. Campbell. It was everyday married life. We lived off post, money was tight. We never lived in post housing; I didn't believe in that. We found a big old home in Tennessee near Ft. Campbell, I'll never forget it. It had been divided into two homes. There was a big board wall right down the middle, the other guy had the front part and we had the back part. It was in an okay neighborhood, not too bad. Schools were okay. We used to take a lot of trips to North Carolina, to visit her mother.

We were married for forty-two years. Late in her life, she suffered from some really bad medical problems. I took care of her until she died, but it just went from bad to worse. She had real bad diabetes, she had COPD. She lived with me on The Farm in Virginia, she passed away here. I took her back to North Carolina to be buried there. She passed away in 2006."

In time, Dutch's adopted daughter Dena had two sons of her own. Dena's family spent several years living with Dutch and Cathy. Dutch developed a very strong relationship with the boys, inspiring them both to lives of selfless service.

"My grandsons are great, we keep in touch. The grandsons are terrific, they're mine. I practically raised them, as they lived with me for so long. Their names are Bradley and Nathan Byrd. Both of them are police officers. Bradley is 44 and Nathan is 42. Both of them did an enlistment in the Army, they deployed to Afghanistan

and Iraq. Both were trained and assigned the MOS of Military Policemen, but they were used as Infantry on those deployments.

Bradley had 22 years in the US Border Patrol, and he's gotten sort of a bad deal. He's got bad lungs, from exposure at the Arizona training center. He was visiting us once, and we brought him a birthday cake. He went to blow out the candles, and all of a sudden, he's coughing up blood. We told him to see a doctor the next day. They kind of hemmed and hawed around, but he went to a doctor off-base. They diagnosed him with some kind of pneumonia. They couldn't even fly back, they had to drive back home. The Border Patrol gave him two months of leave. His symptoms increased, moved into his extremities. He did pretty good with physical therapy. He ultimately resigned from the Border Patrol, and took a job with a Sheriff's Department in South Carolina. He's happy now. His wife is half German. They just bought a beautiful brand-new home. He's a School Resource Officer at a public school. He has a daughter, so I have a great-granddaughter named Makayla. She's been having to do remote learning on computer due to COVID quarantines. It's been hard for her.

Nathan is on the streets as a working police officer. He works for a small department in Maiden, North Carolina. He just made Sergeant, over a whole bunch of more senior people who have been there a lot longer than he has. He's going to college and looking to put in for a Lieutenant's position at his department. He's on the national SWAT Team. He's hard as anything. Before they went in the Army, I used to take them both on rucksack marches, and work out with them to get them fit. He loves it. I just hope he doesn't get hurt. I worry about him, but it is what it is.

Even though they aren't my grandchildren biologically, as their mother is adopted, they are mine. I'm their granddad. They both know that. They'll tell you."

Dutch's pride in and respect for his grandsons is clearly evident when he talks about them. So while they do not share his genetics, his example of public service clearly got passed along to each of these military veterans, now serving as guardians of our civil society. This is one small part of Dutch's enduring legacy.

CHAPTER 19:

CITIZENSHIP!

Dutch left Germany in March 1966 as an E-5 Sergeant, and received orders to the Headquarters Company and then upon volunteering, to the Reconnaissance Company, First Battalion, 506th Parachute Infantry Regiment, 101st Airborne Division at Ft. Campbell, KY. His unit had orders to depart for combat in Vietnam almost immediately. Dutch was soon going to war. But according to US Army policy at the time, he could not do so until he became a US citizen. So Dutch of course completed this important step. As many of his new countrymen had done before, Dutch forswore all prior allegiances, and took upon himself the awesome privileges and responsibilities of American citizenship.

"When I came back to the US in March 1966, I came back to the 101st Airborne Division. I was an E-5 SGT at that time. I became a citizen about a week before departing for Vietnam. I had studied for the citizenship test for about three weeks. I was told to go down to the local courthouse, and see the judge. He gave me the test of a few questions, and said I had passed. He told me to raise my right hand and I took the Oath of Citizenship, and I was officially an American! That was on 30 November 1966. I was only home for eight months. I came home in March and in December, we were gone."

CHAPTER 20:

FIRST TOUR IN VIETNAM. JANUARY 1967 – SEPTEMBER 1968

"I've always been stupid enough that when they asked for volunteers, I raised my hand, before I even knew what the hell it was that I was volunteering for. I got to Vietnam on the 1st of January, 1967. The 101st quickly asked for volunteers to attend extra training – of course my hand went up. I was selected, because I already had some experience. It was a man-tracking school, run by the British Army, held in Malaya. In February, we sent an eight-man team to the school, where we were taught to track down men in a jungle environment. The course was six weeks long. There were classes, then demonstrations, and then we would go out and do jungle patrols, trying to track down instructors. It went fine, I enjoyed it! One instructor there, a New Zealander called Lieutenant Kiwi, really had it in for me. He was tough. It was hellacious training, as hard as SF at that time, six weeks in the jungle environment. I had to order running shoes for my team, running in boots was not going to cut it. It was a good time, I made good friends. I made honor graduate.

After I graduated, I was asked to be an instructor and I stayed on for about a year. I was officially assigned to the 4th Infantry Division during this period, as they were in charge of the slot at the school. When we graduated a class, we instructors would go back to Vietnam with them, and go out with the students on their first few combat missions to advise them and see how they performed in the field. I enjoyed the hell out of that. We were based way out by the Ia Drang Valley in Vietnam. This was very tough territory. That's where most of our missions were. The Battle of the Ia Drang had already happened in November 1965. The Korean Army had later had a unit there, they got kicked out. We had a few contacts with the North Vietnamese during the mantracker missions in the Ia Drang. We bumped up against them, made contact and exchanged fire. Then the enemy took off. These didn't lead to any major engagements. When big engagements did happen, like the base camp being attacked, the trackers would be in the middle of a much larger unit, so we were not in the thick of it. We would find the enemy, and the 4th ID or 101st AB would send out a larger unit to take them on and a big fight would ensue, but we wouldn't be let out with the big force. A couple of times we did get into it. One time I was walking behind a point man holding a tracker dog. His leash had just touched a booby trap wire. I pulled him back – and the wire was right across the dang trail. It was getting dark at that time. That was lucky.

Then we would take a new group back to the school and start again. A man-tracker team most often had eleven members and two dogs, but the dogs didn't work out so well, usually. They would get tired, get frustrated, and every mud puddle they found, they would go roll around in it. So I preferred to do the job by eye. A tracker team would be attached to an infantry unit. Our job was to find and engage the enemy. The regular infantry troops didn't want to do that. So they didn't like us, didn't want us. I didn't blame them, really. I've seen a whole company of soldiers be killed in one day.

We had a little compound of our own, because it was a classified unit (the tracker detachment.) One time a reporter came up and asked us who we were, what we were doing, and he wanted to come into our area. I told him to go away, he wasn't allowed in. He said "I'll go to the General, and get permission." I told him "go right ahead, you do that." He never came back. One good thing happened. One day, the CBS reporter Walter Cronkite and his camera team came by. He wanted to know who we were. I told him "none of your damn business!" He said he needed to know. I told him he didn't need to know a damn thing, he needed to get the hell out of there. I told him, "furthermore, my men are not here to protect YOU. That cameraman, he makes $250 a day, my guys make fifty cents an hour. They are not here to protect your dang life – you better go back that way." This was somewhere out in the Ia Drang Valley. That video made it onto the evening news in 1968! My sister saw the broadcast and recognized me. I never saw it, but my sister did and told me about it!

At the Mantracking School I was put in charge of a platoon of twenty-four Gurkhas who were working at the school as handlers and support staff. They were great people. I learned a lot from them. They were outstanding soldiers. You just had to get used to the way they did things, and their odd ways. Like stopping for tea at 3:00 pm; I didn't do that. One of them would go out and hunt down a bird, and they would eat it, things that no one else would eat at that time. But they were outstanding soldiers; very hard. If you go back through British history they fight all the battles for the British. The British Army asked me, and I took a test to attend the British Commando School, and I wanted to go. But the Army said, "no you're not going to that school, you're going back to Vietnam! I did that and I stayed for another six months after, so I was in Vietnam for 18 months on that tour."

CHAPTER 21:

SPECIAL FORCES SELECTION AND TRAINING

At this point, Dutch took a major step in his life; one with far-reaching consequences. He volunteered for the US Army's Special Forces, the legendary Green Berets.

If you haven't seen it, the 1968 movie *The Green Berets*, partially directed by and starring John Wayne, gives a good depiction of the organization, capabilities and ethos of the Special Forces of the time. While the movie was contemporaneously savaged by the critics due to its pro-military slant and supportive depiction of the SF role in Vietnam, it was popular and ultimately profitable. Although the tactics displayed on film by the Hollywood actors are relatively laughable, the movie remains a Vietnam War classic. It showcases what SF was like at the time, and captures the complex nature of SF missions in Southeast Asia well. It also makes clear how many brave SF soldiers were lost in battle, leaving behind their names and battle decoration legacies on numerous places and facilities. Dutch volunteered during a time when SF losses were extremely high. It was September of 1968.

A classmate of Dutch's in the SF Qualification Course was CPL (later SFC) Melvin McIntyre, who was after graduation assigned to MACV/SOG CCS. McIntyre remembers attending Phase I at Camp

Mackall with Dutch, and also sharing a duplex house with Dutch and his wife Cathy in Eureka Springs, NC. **HE RECALLS:** *"Dutch and I were both pretty much down-to-earth hardcore. 'Fuck the Instructors, they're trying to screw you over, and that ain't happenin'.' We'd just kind of look at each other and half-way smile at 'em, really, and they'd try to pull stuff on us. We supported each other, and it was just like we were kind of like brothers, you know? Us against the world, if you will. Once in training, an instructor looked at Dutch, with his Asian features, and called him a Japanese. That didn't go over well! Dutch was a strong professional soldier who was determined to get through that training, and earn his beret…and also to assist other people to succeed. That's the kind of stuff you did. We got along really good."*

DUTCH RECALLS: *"I got tired of it and put in for SF training. I had met and talked to a couple of SF guys in Vietnam. They wrote letters for me, recommending me, and told me where to go to apply. I put in for it before I left Vietnam and I got selected into SF. I had to take all the tests over again. I made it, no problem. I got back to the US in September 1968 and started training immediately at Ft. Bragg, NC. We bought a house in Fayetteville, NC, right off one of the main boulevards.*

The training was hard, but it was good. We had a First Sergeant, he was a real asshole. He was German; we all called him "Sergeant Shultz" like in 'Hogan's Heroes.' I was already a Staff Sergeant, E-6, so they had put me in charge of a larger platoon of trainees. One night I got called into the company area in the middle of the night. Half of my platoon were Combat Engineers. They had rigged his chair and his desk with demo (demolition explosives.) That wasn't funny. If he had sat down in the chair, he would have blown sky high. So the command said "we've got to search all the lockers." We had an inspection. You wouldn't believe all the demo and detcord we found. They kicked a bunch of them out of the program immediately and sent them back to the 82nd. Involved, not involved, it didn't matter. No questions asked, they were gone.

We had another funny incident. I was driving by the commissary, on the way to my house. I looked over and saw this enormous life-sized gorilla statue, which was an advertising gimmick mounted on a tower by a local radio station on Bragg Boulevard. Several soldiers had stolen it and they were carrying it into the recreation area – they had taken it from the tower in town. Sure enough, within an hour, I was called in – it was my guys. The gorilla was sitting on the pool table. We called the radio station, and they said, "just bring it back, and it won't be a problem." The guys had to steal a stake-bed truck, load it up and return it and then return the truck before morning without getting caught.

For some of us, the actual training wasn't that hard, because we were just back from Vietnam. We already had some experience. Actually, the first part was pretty easy. They put you in teams. They put a guy from the 173rd Airborne Brigade, me and another guy from the 101st Airborne together. We, you know, we had it made. We already knew a few tricks. They had us out there for the land navigation course, things like that. We went out there, stashed our rucksacks and walked the course, then came back and picked up our rucksacks. Actually, it shouldn't happen, but you know, what the hell. At night training, it was kind of cold. People would build fires, when you're not supposed to build fires. We would put ponchos around them, and the instructors were so inebriated, they wouldn't see them. They would drive right by us, hollering and whatever else, and wouldn't even see it. I remember thinking 'boy, if I was in charge of this kind of training, this kind of stuff wouldn't be happening!' not knowing that I was going to take it over, years from then.

The weapons training was the better training, on a wide variety of weapons, and to a high degree of skill. The mortar training was pretty hard, due to the math calculations. I could see why an SF team could hold out against a much larger group of people, like we did in no-man's land, like up in the Ia Drang Valley, which was some of the worst territory that the Vietnamese owned. We had a couple of contacts out there, with my tracker team in the 101st. Like when the Vietnamese attacked that SF camp at Lang Vei. That was bad. (Later on, when I was instructing at the SF School, we had two courses, a Reconnaissance course, and the SF Qualification Course. We used to take them for a long walk, 43-45 miles, from Camp Mackall to the Uwharrie National Forest. On one of these exercises, one of my Assistant Instructors was a black guy named Ron. He told me 'It's time for a break, Dutch'. So we sat them down, and he asked me 'do you want some tea?', knowing that I always carried tea myself. He said 'I've got some real good tea!' He had a slug of vermouth or whatever else in it! He was at Lang Vei. He later died of throat cancer, he didn't even smoke.)

Dutch recalls one humorous incident as a student during Phase I training involving his teammate CPL Melvin McIntyre. *"I was going to write the OPORDER, and Melvin said he would tend the fire. After a while, I thought, "I'd better go check on him." I found him hunched over the fire. He had fallen asleep, and his jacket was smoking! I had to pull him off the fire and shake him awake before he got badly burned.*

We went to the field for Phase III, the practical testing. Unconventional unit leadership, unit decommissioning, things like that helped me a lot in later years.

With regard to training, the biggest thing was just, **stay with it. Don't quit.** *That's what it is, you just don't quit. The kind of people we had in those days that came through, they knew not to quit. A lot of them had combat time."*

Dutch, of course, didn't quit. He earned and first donned his green beret in May of 1969. He immediately received orders to the 5th Special Forces Group in Vietnam. Dutch was going back to the war.

CHAPTER 22:

MACV/SOG

"WAR IS A MERE CONTINUATION OF POLICY BY OTHER MEANS."
— *Carl von Clausewitz*

The history of the Vietnam War is complex, and huge numbers of books and studies exhaustively document it for posterity. This book cannot hope to explain these myriad details accurately or concisely. But any study of the war brings one glaring fact to the reader's consciousness: the war was disastrously affected in conduct and outcome by political considerations. Decisions were made, rules imposed and tactics constrained by politicians, for purely political reasons. A case can be made that the final outcome was doomed from the start, given these realities. The North Vietnamese had been at war with the French from 1949-1954, and were already expert in conducting a guerrilla war of attrition against a larger, more powerful western enemy by the time US forces were committed to supporting the war in Vietnam in 1954. Large scale deployment of US troops began in 1963. One of the most important facts to know of how the war was conducted is that the North Vietnamese and their Russian and Chinese allies used the neighboring nations of Laos and Cambodia as safe havens. They moved massive numbers of troops and enormous quantities of munitions and supplies into and out of those countries with relative impunity, in effect hiding behind imaginary national boundaries on western maps, believing that American forces could not attack them there, as this was the sovereign territory of non-belligerent nations.

American military leaders quickly realized that North Vietnamese forces and materials in Laos and Cambodia would have to be monitored and dealt with, but international law and public perception required that the US appear to be respecting the territorial sovereignty of these countries. Any operations inside Laos and Cambodia would have to be conducted discreetly, with public deniability. So a clandestine war had to be conducted, on a relatively small scale and in maximum secrecy. And for this purpose, the Military Assistance Command Vietnam – Studies and Observations Group (MACV/SOG) was established.

From the USASOC History Office, USARMY.MIL website article dated 25 January 2019:

"The Military Assistance Command, Vietnam, Studies and Observations Group (MACV-SOG) was activated, January 24, 1964, to function as a joint special operations task force. Commanded by a U.S. Army Special Forces colonel, MACV-SOG was a subcomponent of MACV. Born from a need to conduct more effective special operations against North Vietnam, many Central Intelligence Agency programs were transferred to SOG, which eventually consisted of personnel from U.S. Army Special Forces, U.S. Navy Sea-Air-Land (SEALs), U.S. Air Force, U.S. Marine Corps, Force Reconnaissance and CIA personnel. Special operations were conducted in North Vietnam, Laos, Cambodia, and S.outh Vietnam.

MACV-SOG grew in size and scope over the next eight years. Missions evolved over time, and included strategic reconnaissance, direct action, sabotage, personnel recovery, Psychological Operations (PSYOP), counter-intelligence, and bomb damage assessments. Maritime operations covered the coastal areas of North Vietnam. PSYOP missions included 'Voice of Freedom' radio broadcasts into North Vietnam, to publicize the advantages of life in South Vietnam.

The so-called 'Ho Chi Minh Trail,' a vital enemy logistical system named for the North Vietnamese communist leader, was a target of many operations. The trail was a well-developed 'highway' that ran from North Vietnam through Laos and Cambodia. The communist insurgency was sustained by the trail, as troops, trucks, tanks, weapons and ammunition flowed south into South Vietnam. Aerial reconnaissance of the trail was difficult; SOG teams provided the most reliable 'boots on the ground' intelligence.

SOG headquarters remained in Saigon, with subordinate commands and units located in various forward operational bases over the years, with command and control camps, launch sites, training centers, and radio relay sites in all four U.S. Corps Tactical Zones. By late 1967, MACV-SOG had matured and split into three subordinate geographical commands: Command and Control North, Command and Control Central, and Command and Control South. CCN, at Da Nang, was the largest in size and conducted operations in southern Laos and northern Cambodia. CCC, at Kontum, also operated in southern Laos and northern Cambodia. CCS, at Ban Me Thout, was the smallest, and operated in southern Cambodia.

SOG command and control sites operated independently. Each was organized based on the ground tactical situation, but all three had reconnaissance, reaction or exploitation, and company-sized security forces. Each site was about the size of a modern SF battalion. Reaction or exploitation forces were used to extract reconnaissance teams or conduct raids or other assault missions. Reconnaissance Teams (RTs) consisted of two-to-three Americans and six-to-nine indigenous personnel, normally Vietnamese, Montagnards, Cambodians, or ethnic Chinese. Teams were given a variety of code names (U.S. states, poisonous snakes, weapons, tools, or weather effects). Support troops on site provided logistics, signal, medical, and military intelligence support.

Each mission was unique, but most followed a similar tactical profile: after being alerted of a mission, the Reconnaissance Team was briefed and conducted detailed planning, rehearsals, inspections, and training, time permitting. Teams were inserted by helicopter into the target area. Team leaders were Americans and designated as One-Zeros (10), with American assistant team leaders, and radio operators serving as One-Ones (11) or One-Twos (12). Indigenous troops were Zero-Ones (01), Zero-Twos (02), and so forth. Teams were given considerable latitude regarding tactics, uniforms and weapons. Captured enemy equipment was often used. Vital communications were maintained with a Forward Air Control fixed-wing aircraft. Such airplanes coordinated for close air support for immediate extraction if a team was compromised, or upon completion of the mission. A mission lasted from three-to-five days. SOG was all-volunteer, and personnel could leave without prejudice."

At the time, even the existence of MACV/SOG was classified, as were all of its activities. Even Special Forces soldiers assigned to other units were not briefed on its existence and operations. SOG was a highly secretive outfit, throughout its existence and for decades afterwards. At its peak, it had about 2000 members, most of whom were not actively engaged in reconnaissance missions. A total of about 7800 men served in SOG throughout its total history. Only now, over half a century after it was established, is some of SOG's history becoming public knowledge. But a few incontrovertible facts should illustrate the extremely hazardous nature of SOG's missions:

- MACV/SOG Recon casualties exceeded 100% – the highest sustained loss rate of any American unit since the Civil War. But SOG also achieved the highest kill ratio in US military history: 158-to-1 in 1970.

- the Medal of Honor was awarded to nine MACV/SOG men, the Distinguished Service Cross to twenty-three, and SOG members received over 2000 individual awards for heroism. SOG itself was awarded a Presidential Unit Citation in 2001.

- Ten Recon Teams were lost without a trace. Fourteen were overrun or destroyed. Fifty members of SOG are still listed as Missing In Action.

MACV/SOG was disestablished on 01 May 1972 as American involvement in the Vietnam War neared its end.

CHAPTER 23:

MACV/SOG COMMAND AND CONTROL NORTH. SEPTEMBER 1969 – DECEMBER 1970

Dutch's return to combat in Vietnam quickly took an unexpected turn. He was briefly reassigned to the 101st Airborne Division's Mantracker School program, to provide some needed training to the current Instructor cadre. He was there only from June – August 1969. Then he was transferred to the 5th Special Forces Group. When he arrived at 5th SF Group Headquarters in Nha Trang, South Vietnam, he was expecting to be assigned to a conventional SF unit. But 1968 had been an exceptionally costly year for MACV/SOG. Seventy-nine SF soldiers had been killed and far larger numbers wounded. Every Recon man that year was wounded at least once, and about half of them were killed. Replacements were badly needed, and the call went out for volunteers from newly-arrived SF soldiers. As usual, Dutch raised his hand without even fully knowing what he was volunteering for.

DUTCH REMEMBERS: *"After I finished SF training, I went straight back to Vietnam in May 1969. I was about 33 years old at that time. I was always the oldest guy in every school I went to, and I was at that time the oldest guy who had ever finished SF training. I was used to war, and all the things that came with it, so it really didn't bother me that much. It was just normal for me, really. I had spent some time in the so-called Valley of Death, the Ia Drang Valley. But we got out of it. When I went the first time, and then again with SF, it was just a normal time for me.*

When we got to 5th Group Headquarters in Nha Trang, who did I encounter? "SGT Shultz" – the First Sergeant some of the students had tried to blow up at Camp Mackall. They had gotten him out of there right away and sent him to Vietnam. When I got there, they asked for volunteers for SOG. I volunteered, not even really knowing what it was. Then they asked where we wanted to go – CCN, CCC or CCS. I volunteered and was assigned to CC-North, which was a unit that a lot of people didn't want to go to, because they said you wouldn't come back! But I said 'what the heck' and went to CCN and took over a team. It was one of those crazy things. Two or three of us were sitting there and they asked us where we wanted to go: CCS, CCC or CCN. I asked "where is the best place to go?" They said, "South you fight a lower quality enemy. In Central you fight about half and half – tough enemies and not so tough. In the North you fight the really hard ones." So I said 'I'll go to CCN' which was one of the hottest places around because of the nature of the enemy. You had NVA, VC and Chinese fighting you together. But it was no problem. I didn't have any problem with it. I wanted to run recon. I had always been doing recon. So two of us, me and SSG Ted Hornung ended up going to CCN. I was a Staff Sergeant at the time. Ted also became a 1-0 of a Recon Team in SOG. I was there for about a year and a half. He was still there after I left."

Due to his previous combat experience and rank, he was immediately assigned as a 1-0 and told to organize and train a new RT ANACONDA. An earlier RT ANACONDA had been lost, and was presumed to have been killed on a recent recon mission. Their bodies were never recovered. It was a sobering reminder of the exceptionally hazardous nature of SOG missions. While SOG records are incomplete and inconclusive, they do record that on 31 July 1969, a CCN Recon Team led by two Americans was lost and listed as Missing In Action. Of the six-man team, only two indigenous members, named Pan and Comen, survived and were extracted. The two Americans were 1-0 1LT Dennis Paul Neal and 1-1 SP4 Michael Paul Burns. It is likely that this was the leadership of the earlier RT ANACONDA. Both men remain listed as Missing In Action.

DUTCH RECALLS: *"CCN was based in Da Nang, and missions often based out of Quang Tri, about an hour helo flight away. We also flew out of Mai Loc, Khe Sanh and Phu Bai, all of which were even further away. Almost all of our missions were inside North Vietnam or in Laos. I reported in to CCN in May, initially meeting Sergeant Major Hopkins, a great guy. The CO was COL Jack Warren. I was told that RT ANACONDA had been wiped out, and assigned to rebuild a new team with that name. I was immediately made a 1-0 of the new team, so I never acted as a 1-1 or 1-2. I had about three weeks to organize and train them. SGT Steve Hoffman was my 1-1. I ran a total of about a dozen missions as the 1-0 of RT ANACONDA in just over a year, until I got wounded and later departed. For about a year, a lot of the time when we inserted, there were enemy forces waiting for us. One time, I remember we hovered over an LZ with tall elephant grass. The rotor wash parted the grass, and there were NVA soldiers lying there right below us, waiting to ambush us in our LZ. My guys let loose on them and the helicopter took off real quick. It was eventually discovered that there was an NVA officer working inside the SOG S-3 (Operations) section in Saigon. He* *was compromising operations. They got rid of him, but missions were getting worse and worse."*

Dutch's observations concerning the frequent and deadly compromises of SOG missions are consistent with the now-known history of the unit. The North Vietnamese had had decades of opportunity to seed the South Vietnamese Army and Government with spies and infiltrators. Political considerations made informing the South Vietnamese Army Leadership of planned SOG missions a requirement, and operational security was more often than not compromised even before the men left the ground to insert on the mission. In a number of cases, radio intercepts were made of enemy communications, identifying the planned insertion locations, size, mission and individual members of the inbound team to the smallest correct detail. Clearly, highly-classified mission plans were being leaked to the enemy in near-real-time after being given to the South Vietnamese Army. And while a few lower-ranking personnel were discovered or suspected of this, many more operations were compromised than can be laid at their feet alone. Eventually, an American in the SOG OP-35 Operations Section was also found to have been selling mission information. There were undoubtedly many more spies, extending far higher up the South Vietnamese chain of command. The problem was never solved; the American side proposed a complete counterintelligence vetting of all members of the larger chain of command. The South Vietnamese side agreed in principle, but dragged their heels and the polygraph examinations and background investigations never took place. Good men died in huge numbers, and no one was ever found responsible or brought to just accountability for this wholesale treachery.

CHAPTER 24:

SOG MISSIONS

A SOG Recon Team might be assigned any number of a wide variety of missions. These could include Enemy Unit Reconnaissance, Sabotage, Bomb Damage Assessment, Psychological Operations, Communications Intercept, Personnel Recovery, POW Rescue, Sensor Emplacement, and Prisoner Snatches among others. Mission planning followed a fairly standard format and procedure.

"The Intel people from SOG Hqs or The Pentagon would bring us a target package. We went into isolation for about a week, studying photos, sand tables, preparing equipment, etc. It was always a week of isolation." This was for operational Security reasons, to try and prevent leaks of mission destination and plans. Another very important reason, aside from the obvious, for spending a minimum of one week in isolation was to "cleanse" their bodies. American RT members accomplished that by everyone on the mission team eating indigenous rations (Vietnamese food) during the entirety of isolation in order to help their bodies alleviate the unique "American" smell. Urine and excrement odors and consistency directly reflect the food intake that produced them, and RT members did not want any excretions left behind to give clue to the presence of Americans in the patrol area. Additionally, they stopped bathing, or if they did choose to bathe they did not

use American soap or shampoos. Even foot powder was forbidden, given that the issue jungle boots were vented and there was a strong possibility of leaving traces of white powder on a trail or in the jungle. This oversight had proven near fatal to more than one recon team. Those team members that smoked stopped smoking American cigarettes, not that RTs allowed smoking in the field, but to ensure that there was no odorous trace in their clothing, body fluids or respiration. Some teams were not as hard and fast in instituting the above procedures. Those that did were usually a bit more successful than those that did not.

"We would study the target, plan an operation, and rehearse for a week or so. Then we would brief a senior officer about the operation plan. This was usually a joke; they often seemed to be half asleep in our briefings. But we had to get approval, that's just one of those things, the way it was. Then we usually went out on the mission from Quang Tri, but a couple of times we flew out of the 101st Airborne's camp at An Khe, and a couple of times out of Thailand. Those were not too dang friendly to us, as they were Air Force people. There was one time, the Crew Chief tried to kick me out of the aircraft when it was still too high. They were very different from the SOG aircrews we were used to."

SOG Teams were given enormous leeway in the selection of weapons, equipment and uniforms for their missions. Unofficial uniforms were often worn, and a non-standard "tiger stripe" jungle camouflage pattern (featured on the cover of this book) is now widely recognized as a symbol of this unit's operations. As deniability was an important aspect of the clandestine operations they conducted, SOG men used a wide variety of non-standard tools. Team members would sanitize themselves and their uniforms before missions, carrying and wearing nothing which could identify them as Americans on patrol. Sometimes they would wear enemy uniforms and carry enemy weapons, making them blend in to enemy forces in their area of operations. This ruse could often allow them to escape compromise, or to gain an immediate advantage of surprise if they unexpectedly made contact with enemy forces on patrol.

These were small, highly mobile teams, so heavy weapons were not often carried, and some unusual modifications were made to allow heavy firepower to be taken along with minimum possible weight.

DUTCH RECALLS: *"I carried a CAR-15, which everybody carried when we could get them. I often carried a 1911 .45 pistol, and sometimes a suppressed High Standard .22 pistol for sentry or dog elimination, and lots of frag grenades and a few smoke grenades for signaling. Although we had this famous custom SOG knife, I carried a Gerber camping knife; I still have it somewhere. We sometimes carried White Phosphorus ("Willie Pete" or WP) grenades for cache destruction, and Claymore antipersonnel mines. We never carried an M60 machine gun, as it was too heavy. We would make our own "gas grenades" – filling a canteen with ground CS tear gas powder, sealing it up and then putting a small C-4 charge on the outside, with detcord and a clacker initiator. This was like an improvised tear gas grenade, without fragmentation. We made them in the shower, we had to close it off for use because we had CS powder all over us, in our eyes, etc. We had to take our clothes off when making them, as it put CS dust all over us. We carried the PRC-25 long range radio for communications. I carried it, while the 1-1 carried the battery, it was like 35 lbs. Only one or the other of us operated it. We each also carried an ARC-10 radio for short-range comms. We almost always carried a sawed-off M79 grenade launcher, with the sight assembly removed and cut down to a pistol grip, with frag rounds only, no buckshot. One of my guys, a little skinny Cambodian, was out of ammo one time; he chased an enemy down and clubbed him to death with it. I put him in for, and gave him, a Bronze Star medal (this was unusual at the time.) He sold it, he didn't know what it was."*

As you can see, each man would be carrying a formidable personal load of equipment on patrol, often approaching 100 pounds including water, rations, radio, weapons and ammunition. The seasonal weather conditions and jungle and mountain terrain of North Vietnam, Laos and Cambodia made for extreme physical exertion under these loads, even if the enemy was not encountered. But quite frequently, they were, and pitched battle ensued, with the small, vulnerable teams forced to run for their lives, or to hunker down and fight for survival under intense attacks while awaiting extraction. Air support at these times was critical to survival. Carrying these improvised systems and powerful weapons was itself highly dangerous. Dutch recalls: *"We had a medic in SOG, who was carrying a box of ammo, including WP* (incendiary white phosphorus grenades) *once. He dropped it and a WP went off. It got all over him, he was terribly burned. He begged the guys who responded to kill him, but of course they couldn't."*

In any case, a SOG mission "across the fence" in Laos, Cambodia or North Vietnam was inherently extremely hazardous. By design a small, lightly armed team was inserted deep inside enemy territory, on foot. And as mentioned earlier, in many cases, their mission plans had been compromised to the North Vietnamese before they even departed from their operating base. As time went by, the North Vietnamese developed special anti-commando units whose training and mission tasking were to track down and kill the SOG commandos. Often the team made contact immediately upon insertion, or shortly thereafter. And in many cases, only the extreme bravery, exceptional skill at arms and tremendous physical fitness of the team enabled them to survive and be extracted. Another factor which any SOG man will instantly cite is the air support provided to them, often danger close, by valorous and highly-skilled aircrews of the Vietnamese and US Army and Air Force. Another factor was the close support of SOG teams by so-called "Covey Riders" – experienced SOG recon men riding in the passenger seat of light observation aircraft overhead, to coordinate communications and air support for the teams when on the ground. Covey Riders made numerous saves of beleaguered Recon Teams.

In 1970, CCN's RT ALASKA was led by then-SSG (Later CSM) Jim Wheeler. Wheeler is a personal friend of Dutch and the author, who was assigned to CCN in the same Recon Company as Dutch. After starting in the Hawaii National Guard, Wheeler volunteered and earned his Green Beret in March of 1967, and was assigned to the 7th Special Forces Group in what was known as *"Delta Company, the highest ranked company in the US Army." Our CO was LTC Bardis, we had two CSM's, it just ran downhill from there. They just had nowhere else to put us."* Wheeler knew a superior officer in 7th Group, COL Schuengel, who was going back to SOG and invited Wheeler to go to CCN once he arrived in Vietnam. Wheeler arrived at CCN in May 1970 and served there until March 1972. Jim began as RT ASP's 1-1 under SP4 Kenneth R. Wimmer, who had replaced SSG Donald Maley as 1-0 when he was medevac'ed due to a wound. He joined Ken Wimmer prior to attending the multi-week SOG Reconnaissance Team Leader's Course, also called "1-0 School" at Long Thanh, and began running missions as his 1-1 when he returned. After several missions with RT ASP, he was reassigned as the 1-1 of RT ALASKA under his old friend and classmate SGT Robert "Bob" Gulley, who was 1-0 of that team. When Gulley departed, Wheeler took over as RT ALASKA's 1-0.

Wheeler ran five missions with RT ASP making no contact. **WHEELER REMEMBERS:** *"It was either good luck, good recon, or just dry holes. We didn't get into much shit at all. SP4 Ken Wimmer was 1-0. We went on a DMZ mission, going down into bad guy country. We ran into all kinds of stuff on the ground. We actually found a bunch of Russian caches of canned food and medical supplies, and brought a bunch of that stuff out. On the last day of the mission, we also ran into a giant wasp nest. Our point man inadvertently hit it, and they just swarmed the whole team. So we had two Americans and four little guys with 'head shots' – I mean wasp stings all over our heads. We couldn't see. We were basically walking around blind. If we would have got hit, it would have been the end of us. But we were able to get hold* *of Covey (which was an O-2 out of Da Nang.) He picked up our shiny (a ground-to-air signal), and he was able to walk us down to a trail to an LZ. We didn't run into any shit, and he kept on looking and looking, and he put a couple of Spads above us just in case we ran into the enemy and we came out safe!"*

After that mission, Wheeler moved over to RT ALASKA, which at that time ran missions in NVA uniform: *"We used to run NVA, and Bob Gulley was a big blonde-haired kid from somewhere up in Michigan, about six foot two, and he put a pith helmet on his head and a khaki shirt.... yeah, it was a little bit different."*

Environmental risks like wasp nests were hardly the worst dangers SOG Recon Teams faced on each mission. Jim Wheeler's continued experience amply illustrates this. RT ALASKA experienced a harrowing contact on a mission which began on 09 September 1970. The following narrative has never before been recounted in print:

"When Bob Gulley DEROSed, I was then asked to take over RT ALASKA as the 1-0. I had SSG Bob Cedars as my 1-1 and after two missions, SP4 Tom Groark as my 1-2. Cedars was an experienced recon man and sniper. Groark had just finished Recon School and ran his first ground mission with us. We ran a number of missions together, and we ran into a lot of shit. I guess my luck had been used up. On my third operation with RT ALASKA, we got out and first thing on the ground we got hit by trackers, who tried to channelize us, but we went down a mountainside. We got into a creek and ran up the creek and got up the other side to the military crest, staying off the skyline. This was a Laotian target, "Fox Juliet." We ran in six kilometer square boxes, and we could go anywhere in the box, but if you got outside, you had to coordinate, to avoid getting out of "no bomb" lines, etc. They kept pushing us and pushing us, and on the second day, they had finally pushed us to where we had nowhere to go. We were hunkered down and heard them coming down the hill, and we saw uniformed NVA regulars, so we had Sappers, a unit that was designated as anti-recon teams.

The first contact we had was two Chicom hand grenades. So I put my guys in a perimeter, and called a Prairie Fire emergency. Two Chicom grenades and then a Mark 76 US grenade came in. My Interpreter was a guy named Hoang Cha Ly, I'll never forget him. He was an amazing man, he spoke five different languages. He had a beautiful wife, he played about five or six musical instruments, and wrote music, was a tremendous singer. He took all three hand grenades, stuck them up under his chest, and took them, the majority of the impact. Back in those days, all we had was LBE, with a STABO rig hooked up to them (meaning no body armor.) So he was in a bunch of pieces. We all got hit because of the simultaneous explosions. 1-2 Tom Groark, on his first mission, was wounded in the arm and hand. I was unconscious for a while, and when I woke up, I couldn't see anything. I felt my face and there was a mass of blood and meat all over it. I kept on wiping it, and eventually I could see. It was the lung tissue and body tissue from my interpreter. He was over there, just taking his last deep breaths. There was nothing we could do for him. He passed on, and everybody else on the team had been wounded, either by gunshot wounds or by the initial grenade attack. We hung in there for about three and a half hours, fighting these assholes off. Luckily we had put out six claymores; we were able to pop those off and keep them back. And then we got some Spads above us. All they could do was some gun runs, and Willie Pete rockets, some 1.5 rockets around us. I called the stuff in to within thirty feet of us, the gun runs. We were getting a lot of the stuff off of that. I finally got a couple of Cobra gunships out of the 101st, and asked them to squat down right on top of us and just three-sixty their fire. We got Cha Ly (the Interpreter) as much as we could, into a poncho so we could bring all of his parts back. When the first bird came in, I told my 1-1 and 1-2 to get out, with two more "little guys"(only four "strings" came down) and my 1-1 took Cha Ly out with him, and then I stayed. We fought on until the second bird came in and we got out on STABO rigs, and we were outta there, back to Phu Bai, and then Da Nang. When we got up where we could see where all the ordnance had gone off, and all the bodies around us....it was kind of an eerie scene. I say this jokingly now, but I say, when people ask me "have you ever been back to Vietnam?" And I tell them, "almost every night." That was my first Purple Heart."

SSG DON MALEY, 1-0 OF RT ASP DESCRIBES A MISSION HIS TEAM RAN IN MAY 1970: *"I got hurt at the CCN Camp in Da Nang on an engineer stake. I was on guard duty down by the ocean at like three o'clock in the morning, and walking along a wall of sandbags. They gave in and an engineer stake split my left leg open below the knee. It was my turn to lead a mission. The medics said "you can't go on a mission." I told them "look, you don't understand. I'm the commander of this team, I say what I do." "Well, you'd better not," they said. Well, when we got inserted across the line, we jumped off chopper, because we were taking fire when we were going in. I said "we're on the ground, put the other half of the team in, because we can't get out of here now. We're compromised, and it's too late in the day to extract." It was like four o'clock or so when we went in. Anyway, it was a mess, and I split my leg open, and it got infected. The next morning, we got shot out of there, they were up on our ass big time. I went up to take a picture of this trail, and they must have seen me or heard me, I don't know, but all hell broke loose. I was in defilade with my "little people." So we ran back to the rest of the team, and called for extraction, called a "Prairie Fire" emergency. We told them "there was anti-aircraft fire all around here last night" and everything. I told them we were going to E&E* (evade and escape) *and called for a "baseball game"* (code for an E&E effort.) *They finally found us, and the first Jolly Green Giant* (a Sikorsky HH-53 rescue helicopter) *that came out of NKP* (Nakhon Phanom airbase in Thailand) *was going to pick us up, and they got shot right out of the sky in front of me. I was like "Oh, shit, I told you guys I wanted to E&E!" Another one came in, and he got shot out. So the Hueys* (UH-1 helicopters) *came in and one of them got shot down. It was a frigging mess. We finally got out, and when we got back to the launch site at Quang Tri, I told them* what happened and said "the pilots wouldn't listen to me, I told them I was going to E&E, we'll get outta here, and they could come pick us up later." They said "no, we're going to come and get you." Well, they paid for it. That was bad, I didn't like it, but that's what happens. So then I got medevac'ed down to Cam Ranh Bay, because my leg got all infected."

SOG Recon Teams almost invariably were inserted on missions and later extracted by helicopter, often the venerable Sikorsky H-34 Kingbee, flown by Vietnamese pilots of the Vietnamese Air Force 219th Squadron. These incredibly tough machines could and did take enormous damage while getting SOG teams into and out of landing zones under intense fire. The pilots became legendary for their complete bravery and coolness while flying or hovering with enemy fire all around them. Many times, compromised teams deep inside enemy-held territory needed to be extracted in a hurry, with no suitable landing zone available. This might be accomplished through the use of a 90 foot aluminum ladder extended below the helicopter, or of long nylon lines, colloquially known as "strings" suspended from the airframe and dropped down to team members. The string might have a "McGuire Rig" at the end of it; this was nothing more than a large padded loop with an accompanying wrist loop for the rider to use to hold on while in flight. The soldier would sit in the big loop, stick his hand in the smaller loop and signal readiness. The helicopter would then take off, with as many as four men riding below the helicopter at the end of the strings, sometimes for hours to return to base. Sometimes the helos flew through bad weather with the men suspended below them. Later on, after several men were killed by falling from McGuire Rigs, a somewhat safer system was devised. Called a STABO Rig, this was a modified parachute harness, incorporated into the combat equipment harness worn by each man. The harness could be quickly attached to the string rope by a steel carabiner, freeing the soldier from the need to hang on in flight. Still, STABO Rig extraction was highly dangerous and uncomfortable. RT ALASKA 1-0 SSG Jim Wheeler's legendary October 1970 STABO experience is an excellent example of the risks involved.

JIM TELLS HIS STORY THIS WAY: *"I had seven exfils by STABO and I only had one that was really, really messed up. All the rest of them were nut-huggers if you were in them; you could never really get comfortable in them and once you were in the air, there was nothing you could do about it. But it was a lifesaver! We had a lot of guys get drug through wires, get drug through trees, get really banged up. We actually had guys that were drug into trees and hung up, and it brought the helo down...end of mission there. But it was a lifesaver and a means of exfil that the bad guys just couldn't understand. It followed the McGuire Rig. The McGuire was a good rig, but it wasn't as safe. I had no walk-outs. The rest of my missions were aluminum ladder exfils. I much prefer the STABO – less work! Climbing 90 feet up an aluminum ladder while underway is just not much fun.*

The one where I got banged up started off bad. We were going in on a UH-1D Huey at altitude off of the Ho Chi Minh Trail, and we were going to do a wiretap. We were on the ground for less than a day. I was on the second ship in, my 1-1 Bob Cedars was on the first ship in, he was able to clear the ship and we had high grass around the LZ. I slipped down, 'cause we were hovering about fifteen feet above the ground, you could kind of slip down and grab the skid, and then lower yourself and drop down the rest of the way. You're wearing a PRC-25 and close to a hundred pounds of ammo and equipment. I was wearing jungle boots of course and my foot got caught on the outboard gear that leads down to the skid. My foot got lodged in there by the heel and I couldn't get out of it, but I was still holding on to the skid. Thank God I had not let go of the skid. The aircrew thought I was off the skid and they were headed home! Bob just got on the ARC-10 radio and told them "hey, you've still got a guy hanging on underneath!" The Crew Chief finally got out and took a look, which they were supposed to do before taking off. He looked down and there I am frantically trying to get back in the bird, but I couldn't because my foot was stuck. They came back around to the LZ and luckily, it wasn't a hot LZ. I was able to kick loose and drop down. That mission we were doing a wiretap and an Eldest Son (sabotaged ordnance seeding) mission.

Then we went off on mission and as soon as we started moving, we started getting gunfire around us, just single shots – trackers trying to move us in a specific direction and signal which way we were moving. So we had moved about 1500 meters which is a good clip for the terrain up there, and we came across a major trail, and once across it, coming down the hill we heard heavy movement behind us. We found what looked like a knoll off of a cliff, so I told the team to go up and the first guy over the top, my point man went about five steps and took a round through the head. He went down and the rest of us clambered up, and started returning fire. We were able to pick up the so-called sniper, and shoot him, and he was rolling down the hill. Then things got bad and they were running below us, running above us, we were tossing hand grenades. That was the weapon of choice when you were in contact. Luckily they didn't have RPGs. When this happened, it was about five o'clock in the afternoon. The reason we had been interested in the knoll was that it was going to be our night defensive position, but it was so infested with bad guys that I called a Prairie Fire emergency again.

Back in those days, American helicopters were not supposed to be coming back into Laos. The bad guys had helicopters, most of them had old French crews that had gone over to the NVA, and some of them had Russian crews. Anyway, I called in and they said "You're going to have to hunker down for the night." I said "if we hunker down, we're not going to be here in the morning. We are completely surrounded, we are taking heavy fire." And they could hear all of that. So MAJ Ed Lesesne, he was the launch site commander, he said "we're sending choppers out!" and he overrode the rules. He told the US Air Force C130 communications relay aircraft over Laos, call sign "Moonbeam" *"we've got a ground team in dire emergency, with one dead."* (US helicopters were going to be going back into and out of Laos that night, so Moonbeam could ensure that they were not engaged by US forces mistaking them for North Vietnamese helicopters. Moonbeam then broadcast "do not engage" warnings.)

We were right off of the Ho Chi Minh Trail, with double canopy and in some places triple canopy, so depending on what angle they were looking, they couldn't see our strobes. We took one of our LAWs and fired that, and then dropped the strobe into the LAW tube, tied on some 550 cord and a rock and threw that over a tree limb and pulled the LAW tube up high where they could see it shining up, directly over us. I talked them in with the PRC-25, until they were directly over us. I could see the trees moving and I had enough light to see up and see movement and told them to drop the ropes. All of our STABO ropes were on sandbags. Four of them came through, luckily. I only had one American; the rest were all Nungs. So I put four guys on that, leaving myself and my new interpreter as the only ones on the ground. We were still taking fire. My 1-1 had given me all the clackers for the Claymores. I called the second bird over. Only one bag came down. They dropped four, but three got hung up in the trees, so it came down and it was one of the times that the rig we were using was a McGuire on top, with a STABO underneath. So I had my interpreter climb onto my shoulders and clip into the McGuire, and then I clipped into the STABO. We lifted clear and I popped the Claymores and told the pilot to go ahead and unload on the ground as there were no friendlies down there anymore. When they hit that, it looked like they had hit an ammo dump, the place just exploded. We had gotten our dead guy out on the first helicopter. Not all of the Claymores had exploded. When all the stuff went off down there, there had to be a shitload of troops.

Anyway, the flight back was uneventful, other than all our legs went to sleep. The flight was 56 minutes, coming back in from Laos, going back into Phu Bai, which had PSP runways

(Perforated Steel Plate.) Because I had my interpreter sitting on my shoulders, instead of being about a hundred feet below on the STABO (which was normal for one guy) because of the stretch factor, we were hanging about a hundred thirty feet below. They came in at a normal approach altitude, and slammed us into the ground. It was the 11th CAG flying the aircraft. They never even looked. They started dragging me down the runway, and I was able to turn over on my back, and I had my PRC-25 radio on my back, luckily. We were just bouncing all over the place. All of a sudden, my Willie Pete grenade from my belt came off and ignited, and some of my mini-grenades started falling and going off. So the tower (thinking that the aircraft was being fired on) told them to divert to the inactive runway. We couldn't cut ourselves loose. We went about 800 feet down the PSP, then they turned in and went over to the revetments where they parked helos for protection – they were about 16 feet high. They pulled us over one wall, (luckily it was not occupied by a helo,) dropped us in that, drug us over another wall, and dropped us on the other side. We were a mess. That's where my legs got screwed up. They finally realized something was wrong; they were in a hover and the co-pilot looked down through his lower bubble and turned on the lights as there were no more explosions, and realized that we were what was causing the problem. So they cut the lines and flew off and landed. The ground crews got us off of the PSP and the revetments, got us in and got us to the hospital. It had taken all the rubber, the soles and the heels off my jungle boots – completely down. It ground my PRC-25 radio down until it was only about an inch and a half thick. If I hadn't been carrying that it would have been a real bad night. It tore up my knees and my ankles, damaged my back.

After I was recovering, we (The Colonel, myself and entertainer Martha Raye) were able to talk the orthopedic surgeon into not medevac'ing me home, back to CONUS. We got to him and asked him what he really, really needed, and he said he'd like to have a vehicle. So the Colonel ordered the boys to go out and steal a jeep, and they painted it black and put an Embassy number on it, and presented it to him, and the next thing you know, I was free to go!

That's not the end of the story. The pilot wanted to come to the hospital, but our Colonel told him "I don't think you want to go see SSG Wheeler, he's kinda pissed!" But I wasn't really pissed! This guy flew out in the middle of the night, into Laos when he wasn't supposed to. He saved my life – he saved our lives!

Years later, when I was in the 11th SFG as a Sergeant Major down at Ft. A.P. Hill, he came out as an old W4, flying Hueys for STABO and rappel training for the company there. He started talking to a couple of the teams, and said "Every time I get around you SF guys, I have to look around, because I'm scared I'm going to meet this guy!" And he told them the story of dragging this dumb-ass Hawaiian down a runway at Phu Bai. And they said "Oh! That's our Sergeant Major!" That night was a hell of a night, we went to the little club at A.P. Hill and just got drunker than snot together. The minute I was on the ground, I had already forgiven him. It was his Crew Chief and his Door Gunner who would have gotten the worst of it, because they were the ones in back asleep."

DUTCH HAS HIS OWN STORY OF A STABO EXTRACTION'S DIFFICULTIES:

"One time we came back from a mission, we were so deep in country that we couldn't stand to stay in the rig any longer – it was just too painful. I called up and said "You've got to let us down, we just can't stay here any more. We're completely dying." They had to set us down in the abandoned Marine firebase at Khe Sanh. It was completely empty, but full of big shell craters and lots of unexploded ordnance making it very unsafe to land. They had a couple of Cobra gunships fire their ordnance to make explosions in a circular area – blowing us a safe LZ to set down in. They slowly let us down into it and we just sank down into the mud and loose dirt to our lower extremities. We couldn't stand up. They had to come out and help us into the helicopter, we couldn't even stand up."

RT ANACONDA'S 1-1, SGM STEVE HOFFMAN ECHOES THIS SENTIMENT: *"The Hueys arrived with McGuire Rigs but I had on a STABO rig and hooked one of the D-rings into that, but also put my left hand in the safety strap of the McGuire rig. The Huey lifted up with me and three Cambodes and due to the fact that I outweighed them all by 50 pounds my head was where their boots were. The Crew Chief looked and saw the Cambodes were clear of the trees but I was still being dragged through the trees and was pulled a bit off the rig. The backs of my knees were on the McGuire rig seat, and thank God, I had snap-linked into the D-ring and put the McGuire Rig strap on. About 35 minutes later we landed at the abandoned Khe Sanh base and my hand was just white, it was unbelievable."*

CHAPTER 25:

SOG PERSONALITIES

SOG was well-known across its short but distinguished operational history for having numerous legendary personalities serve in it. Some of these men are well-known, due to their having received high battle decorations, or having conducted particularly hair-raising operational feats. SOG recon men were among the finest soldiers any nation has ever produced. Many have been written about in other books on SOG or Vietnam.

One of these men is SGM Billy Waugh. Billy had a legendary SF career and an equally storied multi-decade career in CIA. He has been the subject of two biographic books detailing his exploits. Unfortunately, Dutch and Billy, while mutually respectful to each other, did not get along personally. Dutch was a Staff Sergeant (E-6) when Billy, a Sergeant Major (E-9) arrived at CCN. Dutch already knew Billy by reputation, and was three pay grades junior to him.

"When he came to CCN, he didn't have a job. They had just started a free fall team, and I was on it. Billy was a SGM, and he insisted that he take it over. I was a SSG, and I did not agree with Billy's style or judgment, so I quit that team and went back to my old team. In later years, Billy avoided me when we would be in the same room. Billy got several jumpers killed as Jumpmaster,

dropping them on or too close to objectives. One mission had two brothers on it; one was dropped directly on a guard shack; he was killed immediately. Another time, he sent in a static line team, dropping them in the middle of a river. This was SSG Ted Hornung's RT. They had no flotation gear, and ended up stranded sitting on rocks in the middle of the water. They could have drowned. It was lucky that no one was around and they ultimately got rescued."

Anyone who knows both Billy and Dutch does not find it surprising that they did not naturally mesh as friends. Both men are brave, fearless and accomplished warriors. Both earned numerous battle decorations for incredible feats. Billy's record of combat operations in particular is a litany of risky endeavors, some of which resulted in hair-raising, noteworthy successes. But the two men are polar opposite personalities. While fairly short physically, Billy Waugh is a brash, towering personality, a hyper-aggressive, grab-your-ass, make-it-happen improviser and risk taker. He is forceful, profane and forward-leaning, with a personal confidence noteworthy even among a cadre of extremely accomplished and confident men. Dutch is the antithesis of this persona. Dutch is quiet, thoughtful and thorough. He is likely to fall back on established procedure, historically-proven technique, and sound, prudent planning. While he can be confidently creative when necessary, he is not prone to dependence on aggressive innovation on the spot when things go south due to a bravado-based sketchy mission plan. He is careful, and possessed of a deep well of the quiet confidence of experience tempered by sound judgment. I have met and worked briefly but daily with Billy in a training environment for some weeks circa 1992. I have known Dutch for over thirty years. The two men are simply very different animals, in the author's experience.

Another, lesser known but highly-respected member of SOG was SP4 John Walton. John was the son and heir of Sam Walton, the extremely wealthy founder of one of the world's biggest corporations, WalMart. John Walton was a humble, private man and an excellent poker player. He was considered to be an exceptionally skilled combat medic. He earned a Silver Star when his SOG team was nearly overrun on a compromised mission in August of 1968. He died in the crash of his experimental aircraft in 2005.

John Walton: I knew him. Nobody knew who he was at that time and he didn't tell anyone. He used to go downtown and spend all his salary on buying things for kids, things like that. I couldn't figure out why he didn't need his salary. He had a Silver Star and was an excellent medic.

CSM Jim Wheeler was 1-0 of RT ALASKA, served for a full Army career in Special Forces, and later joined the Central Intelligence Agency, rising to the Senior Intelligence Service before he retired after a distinguished career in CIA. **JIM WHEELER ABOUT DUTCH:** *Dutch was already a legend for having the sit-up record. He could do thousands of consecutive sit-ups. Where I remember Dutch most vividly is, we used to screw with him mercilessly. After he was wounded, he would always make the mistake of sitting behind me in the mess hall. He would be busy talking, and I would shorten up one crutch, so when he would stand up he would fall on his ass. We all thought it was funny. It was hilarious. He just kind of took it in stride. He was a lighthearted guy. None of us expected to live through the war. We would get drunk at the CCN "Recon Club" and somebody would say "let's go hide from Dutch! " And we'd all light out at the same time. Of course, he was the first American at the Malaysian tracker school, so he'd use his skills, he would find us all over the camp! So day after day, we'd just screw with him. As an Operator, Dutch was sharp. He was highly, highly respected. He was legendary, just because of his family and the ordeal they went through in World War II. His knowledge as a soldier was unsurpassed. He was one of what you'd call a "tough guy," but a nice guy. He would always have the time for a new guy, to sit down and share his knowledge. He was just willing to give you all the time in the world to just* *explain things to you. I would love it if I had him at my twelve, my three, my nine or my six. He's just that kind of guy. He's so versatile, you feel comfortable with him on your point, your trail or your flank, as well as leading. The great thing about Dutch is that he can follow, and he can lead.*

DUTCH HAS THIS TO SAY ABOUT OUR FRIEND JIM: *Jim was a big Hawaiian, just a crazy bastard. He was incredibly brave and capable, an excellent soldier. He was a great Recon Team leader. He's a big politician – very social. He knows everybody. Jim's always in good humor. He played a lot of humor with me! I've still got a lot of things to get even with him for. Once when I came tdy to support some training for him in extreme cold weather, I didn't have a sleeping bag with me. He wanted me to accompany the students out on patrol. He said "No problem, I have a sleeping bag for you." What he gave me was a sleeping bag COVER, with no actual bag inside it. I had to sit by the fire all night shivering, I almost froze to death!*

SGM Steve Hoffman led a distinguished Army career. After his assignment as 1-1 of CCN's RT ANACONDA and a brief period out of the service, he served for a total of 30 years. His last assignment was as a Squadron Sergeant Major for SFOD-D, better known as Delta Force.

STEVE HOFFMAN, ABOUT DUTCH: "*I wouldn't be here if it wasn't for Dutch and Don Maley, that's for sure. I didn't know squat about Recon, when we first linked up. I didn't know much. 1-0 school was good, but it was only three weeks. On the first mission with Dutch, I learned more than I did in that three weeks, and then all the missions after, I just kept learning more and more. I think that if I hadn't been wounded, and then when later Dutch got wounded, I would have been able to take over the team. When I came back in the Army in 1971, they sent me to SCUBA school and then in September to 1-0 School, which had been moved back to Ft. Bragg. I made the Honor Graduate there*

– and that is ALL because of Dutch. There's no way I could have done that without having him as a team leader."

MSG Don Maley later served as 1-0 of RT ASP, and retired from the Army at the end of a twenty-year career in August 1982. He was working in the support side of SFOD-D (Delta Force) when he retired. He went to Fayetteville Community College and got an Associate's Degree in Machinist Technology and later went on to a twenty-one year civil service career as an Armorer for Delta Force. **DON MALEY, ABOUT DUTCH:** *"What can I say about Dutch? He was a great man, I thought. He got us out of that mess (the 26 April 1970 mission, to be described later in this book.). He was pointing fingers, doing hand and arm signals, doing this and doing that. He never faltered. I got nothing but the utmost respect for him. You know the old saying "ordinary men do extraordinary things." He did. He was a good man. Everybody knows him."*

SGM Cliff Newman was the 1-0 of RT OHIO, and later made history as a Group Leader in RT FLORIDA; He was the first man off the ramp on that first-ever combat HALO mission, which took place on 28 November 1970.

CLIFF NEWMAN, ABOUT DUTCH: *"The mission on 26 April 1970 was my first foray into combat. I was 1-0 of another Recon Team and my area of operations was weathered in, so we were just kind of sitting around, doing nothing. Dutch came up on the radio, calling in a Prairie Fire, and they were going to launch a Bright Light team for him. They needed a chase medic. I'm not a medic, but I can put on a Band-Aid. So I said "what the heck, I'll go along for the ride." Not thinking much of it, it seemed like Dutch had managed to get out of the contact on his own. I'll be damned if they didn't land the chase helicopter with me on it! I found myself on the ground. We had a brief little shootout, we saw some guys coming over the hill. We managed to get everybody out. I was with Dutch on the Kingbee, a CH-34 that was shot down.*

The pilot did a heck of a job. He auto-rotated to a crash landing. We got out and set up a perimeter with the pilots and crew, we saw people coming over the hill but were able to get out before they caused us any harm. Dutch was at CCN when I got there. He was very well-known, his background was well-known, a very well respected guy. He was a little bit older than the rest of us. He was always very friendly, a very easygoing guy. Obviously very professional. I'm glad to see something come out about Dutch. He's one of those individuals who has gone unsung for his entire career, and his career is exemplary. He's one of very few people who have done the things he's done, and his longevity is mind-boggling."

SSG ANDREW T. BRASSFIELD was also a 1-0 at CCN, and was one of Dutch's closest friends. Born in St. Louis in 1937, he was one of the very few African Americans in SF during the early years of its existence. He arrived at CCN at about the same time as Dutch, and the two became fast friends, attending the same "1-0 School" class at Long Thanh in December 1969. Due to his excellent skills and abilities, Brassfield was selected to lead his own Recon Team. He seems to have had a premonition of death, and he privately told Dutch that he didn't think he would survive his tour at CCN. Unfortunately, this was correct. SSG Andrew Brassfield was KIA on 06 April 1970, leading RT MISSOURI as 1-0 on a mission about 5 miles inside Laos east of Muong Nong, in Savannakhet Province. All the other members of RT MISSOURI were wounded and unable to recover his body as they broke contact and maneuvered for extraction. His remains were never recovered. SSG Brassfield is one of the fallen MACV/SOG heroes to whom this book is dedicated.

ILLUSTRATIONS:

VERSTEEGH FAMILY PORTRAIT, INDONESIA, CIRCA 1936.
FROM L-R: CAROLINA, HER FATHER JOOP WIERENGA, (DUTCH'S GRANDFATHER),
PATRICIA, MILLIE, DUTCH (THE INFANT ON HIS GRANDMOTHER'S LAP)
OMA, AND EMIL.

PASSENGER MANIFEST FOR THE SS RIJNDAM,
DOCUMENTING DUTCH'S ARRIVAL IN NEW YORK ON 14 MARCH 1960.

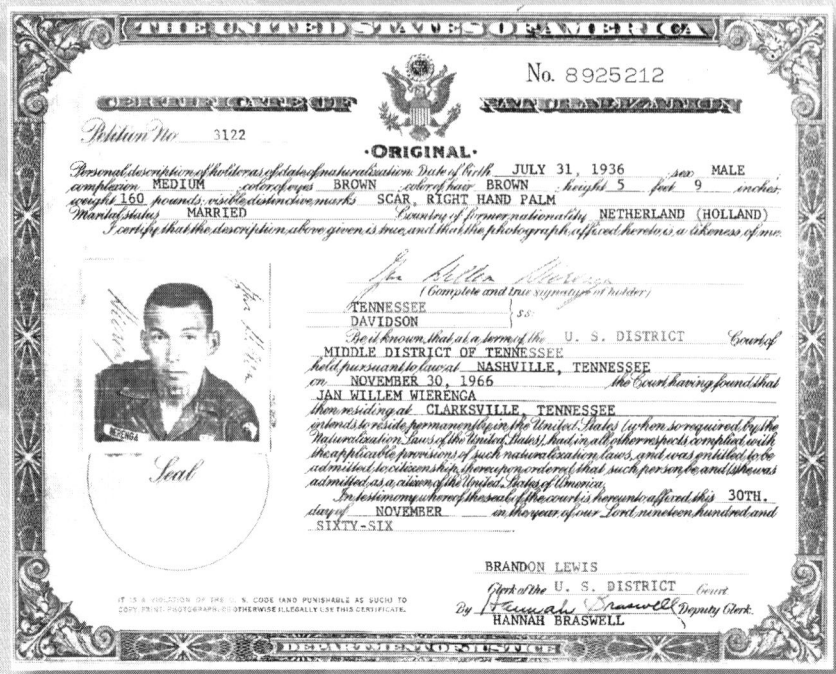

DUTCH'S NATURALIZATION CERTIFICATE, DATED 30 NOVEMBER 1966.

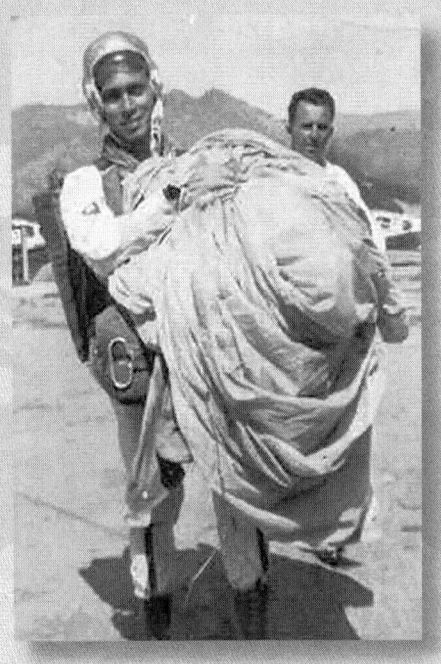

DUTCH JUMPING AT THE FT ORD SKYDIVING CLUB, 31 JULY 1961. THIS WAS HIS FIRST FREE-FALL COMPETITION JUMP. HE TOOK SECOND PLACE FOR ACCURACY.

USNS GENERAL MAURICE ROSE (AP-126). DUTCH TRAVELED TO AND FROM GERMANY ON HIS ASSIGNMENT TO THE 82ND AIRBORNE ABOARD THIS SHIP.

DUTCH AS A SCOUT VEHICLE COMMANDER, SECOND BATTALION, 509TH PARACHUTE INFANTRY REGIMENT, 82ND AIRBORNE DIVISION IN GERMANY, CIRCA 1965.

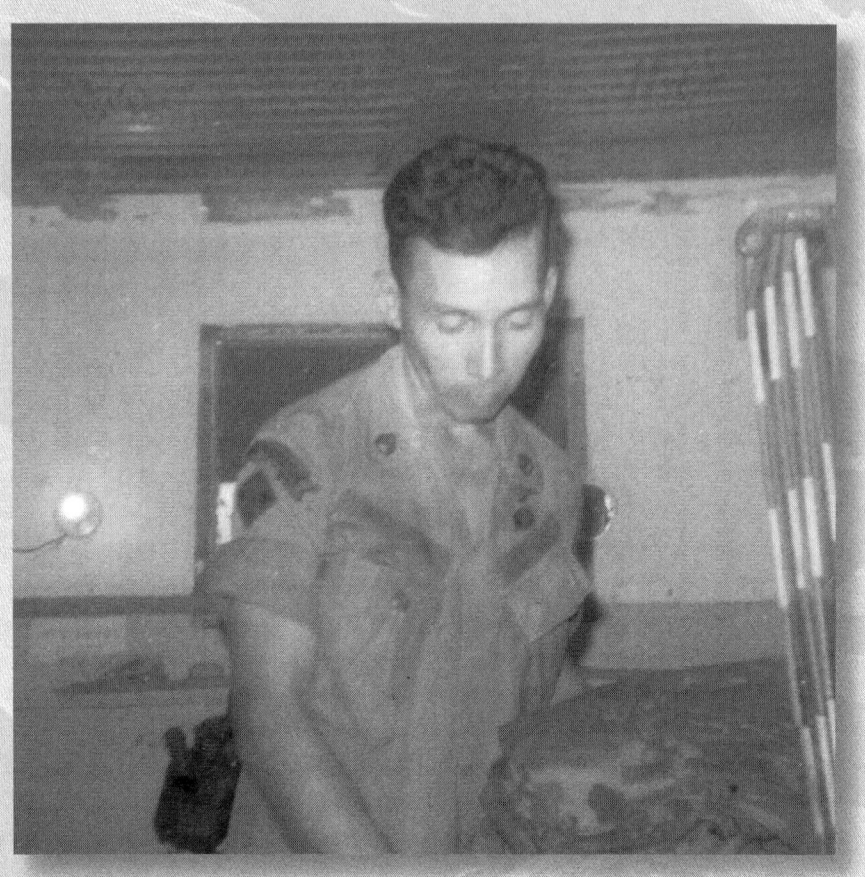

SSG DUTCH WIERENGA, ON HIS FIRST TOUR IN VIETNAM, 1967. NOTE THE 4TH ID COMBAT PATCH AND MANTRACKER SCROLL ON HIS RIGHT SHOULDER.

SSG DUTCH WIERENGA, 1-0 RT ANACONDA, AT CCN FOB-4, DA NANG, VIETNAM, CIRCA 1969.

DUTCH'S "1-0 SCHOOL" CLASS PHOTO, LONG THANH VIETNAM, LATE DECEMBER 1969. DUTCH IS THIRD FROM RIGHT. SSG ANDREW BRASSFIELD IS FIRST ON THE RIGHT, BEHIND DUTCH.

UNMARKED CCN H-34 KINGBEES, ENROUTE TO LAOS TO INSERT A RECON TEAM. (PHOTO COURTESY OF MAJ JOHN PLASTER)

A RECON TEAM BEING EXTRACTED FROM LAOS BY LADDER.
(PHOTO COURTESY OF MAJ JOHN PLASTER)

A SOG RECON TEAM EXTRACTION BY STRINGS UNDERNEATH A UH-1D "HUEY".
(PHOTO COURTESY OF MAJ JOHN PLASTER)

AN A-1 SKYRAIDER OR "SPAD", LOADED FOR BEAR AND READY FOR TAKEOFF TO SUPPORT A SOG RECON TEAM. (PHOTO COURTESY OF MAJ JOHN PLASTER)

A SECTION OF THE HO CHI MINH TRAIL IN LAOS, AFTER A B-52 STRIKE, CIRCA 1970. DESPITE THE HEAVY BOMBING, THE NVA CLEARLY HAVE THE ROAD REOPENED AND FUNCTIONAL. (PHOTO COURTESY OF MAJ JOHN PLASTER)

RT ANACONDA'S UNIT PATCH.
(PHOTO COURTESY OF MAJ JOHN PLASTER)

RT ANACONDA ON MARBLE MOUNTAIN, DA NANG, VN 1969.
L-R: CHAU PENH, SOC RUEN, DUTCH (1-0), CHAU THON, YEND DAY.
(PHOTO COURTESY OF SGM STEVE HOFFMAN)

RT ANACONDA ON MARBLE MOUNTAIN, DA NANG, VN 1969. L-R: CHAU PENH, SOC RUEN, SGT STEVE HOFFMAN (1-1), CHAU THON, YEND DAY. (PHOTO COURTESY OF SGM STEVE HOFFMAN)

RT ANACONDA 0-1 CHAU SOC. SOC WAS BADLY WOUNDED BY THREE GUNSHOTS IN BOTH THIGHS ON 26 APRIL 1970, BUT KILLED NUMEROUS NVA ANTI-RECON TRACKERS IN THE OPENING ENGAGEMENT OF THE FIGHT. DUTCH PRESENTED HIM WITH HIS 1911A1 .45 PISTOL WHEN THE CAMBODIANS WERE DISCHARGED AND THEY PARTED SHORTLY AFTER. (PHOTO COURTESY OF SGM STEVE HOFFMAN.)

KIM TENG, CHAU SOC (O-1), YEND DAY. KIM TENG WAS KIA ON 26 APRIL 1970. O-1 CHAU SOC WAS SEVERELY WOUNDED BUT SAVED BY SGT STEVE HOFFMAN AND SGT DON MALEY. YEND DAY WAS ALSO WOUNDED.
(PHOTO COURTESY OF SGM STEVE HOFFMAN)

YEND DAY AND SGT STEVE HOFFMAN, OUTSIDE THE
RT ANACONDA TEAM HOOCH, FOB-4 DANANG, 1969.
YEND DAY AND HOFFMAN WERE BOTH WOUNDED ON 26 APRIL 1970,
AND BOTH CONTINUED TO FIGHT DESPITE THEIR SERIOUS WOUNDS.
(PHOTO COURTESY OF SGM STEVE HOFFMAN)

DUTCH RECEIVING THE SILVER STAR, BRONZE STAR AND PURPLE HEART FROM GEN CREIGHTON ABRAMS, COMMANDING GENERAL MILITARY ASSISTANCE COMMAND, VIETNAM, AT FOB-4 DA NANG, CIRCA SEPTEMBER 1970. NOTE DUTCH'S CRUTCH, IN USE DUE TO HIS WOUNDS OF 11 AUGUST 1970.

(L-R: SGT DON MALEY, 1-0 RT ASP. SFC CHARLES BOOKOUT, 1-0 RT COLORADO. SSG DUTCH WIERENGA, 1-0 RT ANACONDA. AT THE CCN FOB-4 "RECON CLUB", MAY 1970. SFC BOOKOUT WAS KIA TWO MONTHS LATER ON 04 JULY 1970, LEADING RT COLORADO ON A PATROL IN LAOS. HIS REMAINS WERE NOT RECOVERED.) THE BANDAGE ON MALEY'S LEG IS THE WOUND DISCUSSED IN CHAPTER 24.

MACV/SOG CCN RECON MEN AT PLAY, FOB-4 DA NANG, 1970.
L-R: MERVIN LIBBY, ROIS BLACK, JOHN HOBBS, RAY SLADE,
JOE PRICE AND DUTCH WIERENGA IN FRONT.
(PHOTO COURTESY OF MACV/SOG UNIT HISTORIAN JASON HARDY)

SP4 JOHN WALTON, 1-1 RT LOUISANA AFTER RECEIVING THE SILVER STAR.
(PHOTO COURTESY OF MAJ JOHN PLASTER.)

DUTCH AS A RECON INSTRUCTOR, JFK SPECIAL WARFARE SCHOOL, CIRCA 1972. HE IS DISCUSSING THE FINER POINTS OF TRACKING WITH THE STUDENTS.

DUTCH AS A HALO INSTRUCTOR, CIRCA 1975.

DUTCH'S GRANDSONS, BRADLEY AND NATHAN BYRD, WHILE LIVING WITH DUTCH CIRCA 1987. WONDER WHO HAS BEEN INFLUENCING THEM? BOTH WOULD GROW UP TO BE SOLDIERS AND POLICE OFFICERS.

DUTCH AS AN INSTRUCTOR AT "THE FARM", CIRCA 2001.

DUTCH, KATHY AND GREAT-GRANDDAUGHTER MAKAYLA.

CHAPTER 26:

RT ANACONDA'S MISSIONS

About three weeks after completing 1-0 School in Long Thanh and returning to CCN, Dutch had formed and trained a new RT ANACONDA, assisted by his 1-1 SGT (later SGM) Steve Hoffman. They were joined by SSG Don Maley as 1-2 for one mission in late April. SFC Rocky Kamai ran one mission with them as 1-2 before taking over his own team as 1-0. Later, SSG Roger Teeter, a strapping, 6'4" 240 lb soldier who moved swiftly and gracefully despite his bulk, also ran some missions as 1-2 of RT ANACONDA. He was moved to 1-1 of another Recon Team, and ultimately was KIA while flying as a Covey Rider. In late April 1970, SSG Keith Kincaid took over as 1-1, replacing Steve Hoffman. Kinkaid also later served as a Covey Rider.

DUTCH REMEMBERS: *"Of course we had Cambodians at that time. We started the team with Cambodians. They were a different kind of people. Very religious, if you can believe it. They didn't drink, didn't curse. They would put this kind of a rosary on you for you to wear, take it off you if you were going to the latrine, then put it back on you when you returned. But they were good fighters. I argued with them a few times, like when they wanted a team of seven of us to take on a whole company. I told them we would coordinate airstrikes, and bomb the enemy instead. After the airstrikes, they would realize that we had done the right thing to not go out there and fight an overwhelming force directly. But that's the way they are: they don't care how many men they have. When they fight, they fight. It was pretty good, except for one time, we had one guy who was kinda different from everyone else. You know Cambodians are usually kinda short. We had one guy who was over 5'9", almost six foot tall. One day I came out of my hooch, and this guy had started a fight with one of my Cambodian team leaders. People asked me "and you stepped into the middle of it?" Well, what could I do? That was crazy. They said "You could have gotten killed!" Oh, well. But come to find out, he had been trying to get the team to take us out and lead us into the bad guys. That was what the fight was about. So we chaperoned him out of the gate and turned him over to the 'White Mice' (the South Vietnam police wore white uniforms and usually ran from any fight, so they were known derisively among American troops as "White Mice.") I don't know what happened to him after that. This wasn't common, it is the only occurrence I know of of having a traitor in a SOG Recon Team. Other than that, we had a pretty darn good team."*

STEVE HOFFMAN REMEMBERS: *"I rode up to CCN from Nha Trang in the same airplane with Dutch on 05 December 1969. That's where we met for the first time. I had no experience whatsoever. I had been on an A-Team in 10th Group, over in Europe. If we had to tell about experience in Asia, I could talk for about thirty seconds...Dutch could talk for hours. I remember* *him talking about being on the Combat Tracker team. The next day, we flew down to Long Thanh to go to 1-0 school together. That was a three-week school. When we came back, I found out that I was going to be the 1-1 of the new RT ANACONDA."*

The Team now consisted of these two Americans, and eight ethnic Cambodian soldiers. The 0-1, or Leader of the Cambodian troops was Chau Soc Rund, and the others included Kim Teng, Yend Day, Chau Thonh, and Chau Phenh. Weapons issuance and zeroing, Patrol Standard Operating Procedures, tactics and communications had all been worked out, and Immediate Action Drills repeatedly rehearsed. Team members had gotten to know each other and roles had been assigned and refined. They were ready, and it was time for RT ANACONDA to go operational. They didn't have to wait long.

DUTCH: *"Our first mission was to go and look for the members of the earlier RT ANACONDA which had been wiped out. We went back to where they had inserted, but they were all gone, we found no trace of them. Some of their gear was later seen in a museum in North Vietnam.*

Missions lasted from five to ten days deep in enemy territory. The longest I remember was ten days. One time was only about four and a half days; we had to get out as we had been compromised. My RT was inserted and patrolled to a clifftop site between two NVA Battalion Headquarters. We found the wire connecting the two Hqs communications, tapped into it and were recording the signals intelligence that flowed between them. We were in an underground hide. It went good for a few days. The last morning, one of my team was on sentry, but had fallen asleep. One of the enemy perimeter sentries walked through and my guy woke up and made eye contact. The sentry just gestured in greeting and kept walking, but he immediately reported us. I cut the wires and we moved up the hill to report and await extraction, knowing we had been compromised. The terrain was so steep we couldn't go downhill. We just moved uphill. A company or more was approaching us. They sent trackers after us. I told my guys, "just sit completely still – do not move at all." I called in helicopters to support us. They sent three or four armed attack helicopters, with two guns on each side. They slaughtered them. There was heavy fire all around us as they strafed them below us on the hill. Then they dropped us STABO lines from two helos and we got out of there. It was actually kind of funny. The guys on the lines under the other helo got caught over a tree limb; the pilot kept trying to lift off, but couldn't go anywhere as they went back and forth, trying to get clear, with the rope sawing at the limb. The guys' eyes were as big as saucers as they were hanging on that damn rope. It was really funny. I told my aircrew to tell the other pilot, go forward, and then straight up. He did that and they got out of there. It wasn't the first time I had seen a STABO rig get hung up. Once we had a POW we had captured and were extracting, and the rig got tangled in a tree. They had to cut the rope to free the bird; it was just one of those things out there.

When we sent the tapes we had recorded to the NSA, we asked what was on it. They told us we had no need to know."

1-1 Steve Hoffman remembers this mission well, as he was the person trained to operate the wiretap equipment. **HE RECALLS:** " For this mission it was Dutch and I and six Cambodes. The weather was very bad and we spent eight days at the launch site at Quang Tri before we were inserted. The Area was Golf Eight. That is supposed to have been the farthest north mission ever run by CCN. I did not know what to expect when we got to the wire but it was nothing that I ever dreamed possible in this jungle. I was expecting the thin black wire like used with the Squad Telephone. Instead, there were state-size poles cut to ensure the wire could not be seen above the tree line . Each pole had a lateral with four 1/8 inch copper wires with stateside insulators on each lateral, and two 1/8 inch copper wires with insulators attached to the actual vertical poles. The wiretap had been completed, in the early afternoon of the fourth day. We were supposed to be there five days, but we ran out of tape. We had four cassettes of recorded tape. Dutch called in for permission to exfil. The request for exfil went from the launch site at Quang Tri back to CCN. The CCN commander, a Colonel, said "no, conduct an area recon." Dutch was absolutely stunned that they would risk the four cassettes of tape to do an area recon. Dutch asked for confirmation of that from the Commander, and the answer came back, "yes, conduct an area recon." So we packed our stuff up and we started moving up the very steep hill, and after about a hundred yards we heard voices down at the tap site. From that point, we just hunkered down, laying down, and two guys came up the hill (not on our tracks, because Dutch was very hard core about covering our trail.) We saw them, and they came up and looked around, not really tactically, and they just kind of shrugged their shoulders and walked back down and we could hear them talking and laughing. That was when Dutch requested the exfil, because we may or may not have been compromised at that time. The request for the exfil was then approved. The Covey started coming in, low, but not super low, and they were flying by and that whole mountain just erupted with 12.7mm anti-aircraft fire. I was later told that there were over twenty gun positions on that mountain, covering the commo wire. To get us out, it ended up that we had two Huey slicks and we came out on strings, and there were also some Marine Charlie-model Huey gunships. They had a Checkerboard painted on the front of their bird, and were very good at what they did! They kept everybody away from us on the top of the hill. There was also napalm used by A1E Skyraiders, and there was plenty of it. One of the first places Dutch asked for them to hit was down at the tap site where the trackers were. We were laying there for about three hours for the gunships to suppress the fire."

Another mission was typical of the low profile, but high-impact missions which SOG teams were tasked to perform under extremely dangerous conditions far behind enemy lines.

DUTCH RECALLS: *"One memorable mission, I was tasked to determine if a specific North Vietnamese General was at a particular mountaintop Division Headquarters. It took us three days to get there, from insertion then carrying our rucksacks and equipment, crawling through dense vegetation and heavy jungle for three days, straight up a steep mountain. Once there, I learned how to take long range photos with basically nothing. I aimed my small Olympus Pen-EE camera through my binoculars. I was finally able to determine that there was a horse there. This General was known to be a horse riding enthusiast; he liked to ride every day. I was able to see the horse, and an enlisted man guarding it and tending to it. I also saw a sizable security guard detail, just sort of hanging around the area. I reported this in, and we all took it as a sign that the General was indeed there. I called in a B52 airstrike. We never knew for sure what happened to him after that, but they bombed the heck out of the area. It's unlikely that he survived, as it was a very intense bombing. We walked back out to be extracted after that mission."*

STEVE HOFFMAN REMEMBERS THAT MISSION AS WELL: *"On our second mission, we were in the DMZ area. SFC Rocky Kamai was with us as 1-2. He had had a previous tour in SOG, but he had just gotten back and they sent him out with us to get re-acclimated. Dutch was watching through binoculars and we saw a hooch, way down in the valley, about 500 yards or so. When I say a hooch, it wasn't like a bamboo hooch; it was a place to stay for a long period of time. There was a bunch of people around it, and the horse, and he called an air strike on it. If they told Dutch that there was a General, it did not get down to me, but there definitely was a horse, and it definitely died, and probably everybody around it. The hooch went down too. When we got back from that mission, we were in isolation at Recon Company for debrief, and SGM Hobbs told Dutch and I "you gotta go down to Saigon and get debriefed." So we went down to Saigon and signed into House 10 (a SOG safehouse in the capital) and they told us that we were just there for an R&R!"*

Occasional SOG prisoner snatches were looked upon as glamorous achievements, as they got a lot of attention up the chain of command, and valuable intelligence could be obtained from the captured prisoners. Unfortunately, Dutch's team never pulled one of these highly-prized successes off. *"We tried for prisoner snatch missions a couple of times, but didn't get anyone."* It's too bad they had to cut the extraction rope on that one occasion mentioned above.

1-1 Steve Hoffman recalls another of RT ANACONDA's missions under Dutch's leadership, which took place in late March 1970. *"Our third mission was your basic Area Recon except for the overabundance of excitement. We had to fly to Nakhon Phanom airfield as the target was too far of a flight for Hueys. We were limited to 6 people and we had a new guy as a 1-2 who was going to be a 1-1 on another team. This was SSG Roger Teeter and he fit in well with us. This was a southern Laos mission, in fact the most southern mission we ever did. It was much, much hotter, and much more open terrain, and we ran out of water on the third day. We were in bad shape. When you are thirsty, you just get stupid. We just kind of slowed down and waited for it to get dark, and we found a trail. It was a fairly used trail, and Dutch said "we've just gotta follow this and get water." It was really pitch black dark. You know how you can just tell that there is something beside you? I hadn't really touched it yet, and I thought it was a tree. So I just leaned up against it and it was FLAT. What we had done was walked into a Base Camp. Thank God we did it at night, when they were moving. It was a Base Camp or Way Station for bringing soldiers down into Cambodia and onward, so there was no one there. There were about twenty or so hooches. Dutch said "there's got to be water here." We found it, a nice, clear spring, and everyone drank and filled their canteens and we moved off for 100 meters into a thicket. We woke up early and again watered, to replace what we drank overnight, and then moved up a heavily vegetated finger to watch if anybody was going to move into the Way Station. We were on the other side of a different stream. About forty-five minutes later maybe 150 NVA moved into the Way Station.*

Dutch loved calling in air strikes, I can tell you that. Dutch called Covey and that began about a two-hour air attack on the Way Station initially involving an OV-10 marking the site and four A1E SkyRaiders working in pairs. On their first pass two A1Es hit the Station with napalm while the second pair covered them with machine gun fire. The second pair became the new first pair and again the station was hit by nape covered by the machine gun fire. The next pass was made like the first two but with each pair of A1Es using Cluster Bomb Units (CBU) and covering themselves with their own machine gun fire. When the CBU containers burst there were many of what looked like balls drop, and I thought it was a misfire but seconds later they burst open and White Phosphorus covered the entire area of the Way Station. This was repeated by the second pair until they had expended all their munitions. An unusual thing happened during the bombing; that was suddenly, a thin straight line of fire rapidly started up the side of the ridge on the opposite side of the Way Station from us. We never heard a viable explanation as to what that was. There were screams heard into the night. As we were not attacked, we assumed that we were not compromised but after dark we heard movement from the direction we came from and Dutch called in for air support. The request was filled and shortly after about 23:00 a C-119 Stinger (a potent gunship armed with two six-barrel Vulcan 20mm cannons and four thirty-caliber gatling miniguns) arrived. They pretty much suppressed everything, and we were pulled out late morning the next day. Roger Teeter was a good and brave teammate and moved into a 1-1 slot on another team after the mission. He was killed six months later while serving as a Covey Rider. God bless SSG Roger Teeter."

Under the repeated stresses of these high-intensity missions, men needed to relax and blow off steam when they could. And SOG men occasionally became (or were naturally) fairly wild and high-strung. Wartime tends to lower personal inhibitions, to put it mildly.

DUTCH REMEMBERS: *"After every mission, we had a week off. You could either go to Thailand and stay in a hotel and spend a lot of money, or you could go to Saigon and stay in our hotel, where beer was twenty-five cents in our clubs. We went to Thailand once- that wasn't too dang great, really, it was expensive as hell. We went to Saigon mostly. There, on one of these trips, one of SOG's rear echelon guys, the First Sergeant of the SOG compound in Saigon, a MSG Keene, wanted to take us out in a jeep and show us his favorite Chinese restaurant. We went, and it took over half an hour for our food to come. This made him really mad, and he threatened to shoot out an overhead light, yelling at the staff. We told him to calm down, but he shot out one of the lights overhead. That was his return ticket to the States; he was gone immediately."*

CHAPTER 27:

AMBUSH! THE SILVER STAR. 26 APRIL 1970

The nature of these types of deep insertion missions inside enemy-held territory guaranteed that sooner or later, any RT would experience major casualties. RT ANACONDA's luck ran out that spring.

"We got ambushed on one mission into Laos. This was on April the 26th of 1970. One of the other men on this mission was my new 1-2, who had joined the team not long before. His name was SSG Don Maley.

Don has always been a big bravado kind of guy. He is still alive, living in New Mexico. He was at the last SOG convention. He's had one ear cut off due to some medical condition. Apparently the doctor told him it needed to come off. He has grown long hair over it to cover it. He was one of those crazy bastards you hear about. One time somebody wanted to eat something out of a can. They asked him, "Don, do you have something to open this can?" He said "yeah", locked and loaded a round in an RPD machine gun and shot the can open on the spot, inside the hooch. "OK, the can is open."

Unfortunately, on this one mission, we had three wounded and one killed. It just happened in no time. Our mission was to go out and find the enemy. Well, we found 'em. But we couldn't go anywhere, because they were around us on three sides. It was just by luck, nothing else but luck. I let off a couple of full magazines out of my weapon, and killed a machine gunner, who had just killed my point man in the opening engagement. But in the meantime he got four of my guys – half the team. My Assistant Team Leader (1-1) SGT Steve Hoffman got hit. He had shrapnel wounds in his legs, holes all in his trousers. The guy who was with him, and the guy who was with me – four out of nine were killed or wounded. My 1-2 that day was SSG Don Maley. The Point Man, a Cambodian, was hit through the femur immediately. He was right in front of me when he was shot. He lived for about nine hours before he died. We had a nine-hour firefight waiting for extraction. For a while I couldn't find my 1-1 and the Cambodian team leader. Steve had just gone off in some direction to scout. I finally got back with him. Another machine gunner had us pinned down. Steve had a pretty strong arm, so I told him to throw a grenade at the machine gunner, who was pretty far away. Well, he drew way back and to throw it on a high arc, and tossed that damn grenade. It hit a tree limb and dropped right back down between two of my guys. It exploded right there. But nobody got hurt. It was a hell of a fight. I had to call in air support, a napalm strike. It was so close that it burned the bottom of our boots, but the enemy were crispy critters. We were on top of a little hill and they were below us. A flight of three helos responded to my call for air support and dropped the napalm danger close. They got there just in time.

We got out of it 'cause Cliff Newman came and picked us up. He was with a brand new Bright Light team of about nine or ten guys. They came and got us out. I finally got what I thought was everybody into the helicopter. When they took off, I looked down and saw one of my Cambodian guys, running around like a chicken in a field. I pointed him out to the aircrew and said he had to be extracted. Another pilot, flying a Cobra, landed and picked him up, having him sit on a skid and hang on to a pylon. He flew all the way back into South Vietnam like that, sitting on the skid outside the aircraft. The Pilot's Squadron Commander was going to reprimand him for that – for saving a guy's life.

Back in my bird, I told my guy to tell the Crew Chief "don't go over that mountain." We had been lying there all night, watching tracers crisscross the sky from machine guns on the mountain. Well, they did anyway. He told 'em, and guess where they went... right over that mountain. And that was the end of our helicopter flight, because they shot us down. Well, we all got out of it. Luckily, it was the lead pilot of the Vietnamese Army flying us. He set down on the rocky shore of a big riverbed. When we hit the ground, it was hard. The wheels, normally on struts far below the aircraft, were even with the cockpit after the crash. But we all came out of that one. We finally got another helicopter that was half empty. I remember the Bright Light guys throwing me in that darn thing. I was wearing a big rucksack on my back, all loaded out. They just threw me in like so much weight. We got out of that pretty well, after all the problems we had had the day before. But it was one of the worst missions I ever had. I've never had anything like that before. We took the guy that died... we carried him all the way from CCN to the Cambodian border. We gave his family the death benefit money and turned him over to them. Then we went back to Da Nang. Not long after that, we had some kind of policy change. I had the Cambodians go back to their country. The Team Leader.... I gave him my .45 pistol when we said goodbye. He was really great."

STEVE HOFFMAN REMEMBERS THE EVENTS OF THIS FATEFUL DAY: *"My last mission was less than 24 hours long but filled with misery. We had another new 1-2, SSG Don Maley, who fit in with us very well. We inserted at about 9:00 AM on a hilltop for a POW snatch mission. The reason that was done was that our CCN Colonel was kind of goofy. One team had made a* *successful prisoner snatch, so he wanted us to go do another one three days later. We figured that if we inserted early and moved about 100-150 meters from the LZ site, and set up our ambush, we could catch the Counter Recon Team by surprise and hurt them badly later that afternoon as they tracked us from the LZ. It made sense that if we had a good violent ambush, we could exfil using the infil LZ. Things went well the first two hours as we moved about 200 yards from the infil LZ and set up our ambush, we all had gas masks and were planning on using powdered CN gas taped to our claymores. Everything was set up by 11:30 AM and we settled in and waited for their CRT. It did not come. At about 9:00 PM Dutch decided we should do 50% awake and 50% sleep which was a good call. He planned to gather up all the ambush items and move to another location and set up again.*

By 6:30 AM we had everything packed up when Chau Soc, our 0-1 heard something and began very slowly walking down the hill we were on. As I was beside Soc I got up and followed him. We got about forty yards out, maybe fifty, and Soc just froze. I don't know who shot first, but there were about eighteen people right there. That was the standard for the "following teams" – the Recon Hunters. That little Cambode took down three of them before he was hit two times in the left thigh and one time in the right thigh. I got off two shots and hit the dirt as what seemed like eight or nine rounds destroyed the bush beside me. I was wounded in the right thigh by a ChiCom grenade. I was able to drag Soc to a place where we were in defilade. Our guys could not see the bad guys and the bad guys could not see them which saved the remaining seven of us. The way the CRT members deployed made it seem as they did not see the remainder of our team and they began to move around Soc's and my right. At that point they had moved and exposed themselves to Dutch and the rest of the team. Dutch, Don, and the Cambodes dropped several of them quickly. Yend Day and Kim Teng were the next hit, Yend Day in the right hand and shoulder and Kim Teng in the right thigh. Yend Day began shooting with his left hand but Kim Teng was hit in the bottom of his femur and shards cut the femoral artery very close to the knee. What was not known was that shards also cut the femoral artery much higher up than was thought and Kim Teng died. Chau Thonh killed two CRT members when they tried to get around the right flank and I killed another with a grenade. Soc killed another when the three-man group near us began withdrawing. Everybody was shooting. I heard Don Maley yell to me. He always called me "Dustin Hoffman" – like the movie star. I heard Don yell "I'm comin' for you, Dustin!"(laughing). That was the last thing I expected to hear, but that's what he did. He came and helped both of us to rejoin the rest of the team. I didn't see Dutch at that point, but there's no doubt in my mind that he was doing exactly what he should have done.

The surviving members of the CRT began withdrawing from the contact. We also were in bad shape as we had 50% casualties. The Bright Light team that came in to get us had about seven to eight guys on it, and three of them were from the SF Team I was on at Ft Devens in the 10th SFG! One of them was SFC Gonzalez. We now had Hueys and H34 Kingbees. The wounded including me were put on a Kingbee. I didn't know anything about it, because everybody was speaking Vietnamese on that bird, but we ended up going down and picking up one of the pilots from the crashed Kingbee. I later learned from one of the members of the Bright Light Team who were inserted to assist our four unwounded in getting us out of the ambush site and to the LZ that a A1E SkyRaider saw six NVA, about 500 meters from the ambush site, of whom several appeared to be injured. They were eliminated by a direct hit with a napalm bomb.

When we got back from that, the Kingbees were so low on gas that they couldn't make it to the hospital and they had to land at the launch site at Quang Tri and they sent ambulances to take us over to the hospital. They were afraid that they would fall out of the sky before making it to the hospital. I was medevac'd from the 95th Evacuation Hospital in Da Nang, on May 13 to Camp Drake

Japan. I returned to the states on May 23rd and was discharged from the Army in August at the Valley Forge General Hospital."

DON MALEY'S COMMENTS: *I was on the CCN Hatchet Force for six months, and then after that I said "I'll try Recon Company." So they put me on Dutch's team. I only ran that one mission with RT ANACONDA, because after that we had to dispose of the team. That was the last Cambodian team at CCN. One of the guys got killed. I was trying to keep him alive, but I couldn't do that with bullets flying all around. My equipment was getting hit, but nothing blew up, so I lucked out there. Bottom line, I had to crawl out and get Steve, and one of the other "little people", and drag them back. The Spike Team came in and got us out, that was my old Hatchet Force. Finally the choppers got up there, (we were at high elevation in Laos) and a bunch of us jumped on one of the Kingbees (the H34 with the Vietnamese pilots. They were great pilots.) We were on a hillside, so we told them, "do NOT go down to the bottom of the valley to pick up airspeed and then come up over the other ridge line. Last night there was tons of anti-aircraft fire over there." Well, he does EXACTLY what we told him not to do, so we get shot down. Dutch turned around and looked at me and said "well, we're gonna crash and burn." I said "No, we've got forward airspeed, Dutch, we're gonna autorotate in, don't worry about it, just hang onto something!" So we did that, and the hordes came over that ridgeline after us. We still had our air packet (close air support) with us and they put some fire down for us. One of our "little people" grabbed a machine gun from the door gunner's position, because that guy didn't know what to do or something, I don't remember. He put down a base of fire around the chopper. The last guy on the ground was one of our "little people". We always taught, the last man on the ground, when you get on the chopper, pop a red smoke. Then the air packet would come in and strafe the area, and blow it up. Well, he popped a red smoke, he's still on the ground! Well, one of the Cobra pilots, Dutch called him on the radio, and he came in and lowered the front ammo door on the left side of the chopper,* *and snap-linked him in there and took off. We cleared out of the area, and we still had the air packet. We had left all our heavy equipment in the crashed chopper, that shit was just too heavy to carry. So one of the slow-movers, one of the prop-driven jobs, (the A-1 Skyraider, also often called a Spad) came in and dropped napalm, 'cause all the enemy had gathered around the chopper, to see what was in there. They made crispy critters out of 'em. I flew the body of the one guy who had died in my arms down to the Cambodian border, then we came back and they put me in charge as 1-0 of RT ASP. "*

In recognition of his outstanding leadership and performance in this operation, Dutch was awarded the Silver Star medal on 07 August 1970. His Silver Star award citation reads as follows:

"For gallantry in action while engaged in military operations involving conflict with an armed hostile force in the Republic of Vietnam: Staff Sergeant Wierenga distinguished himself while serving as team leader of a nine man reconnaissance team deep within enemy territory. While Sergeant Wierenga's team was moving from one area to another, enemy movement was detected. Two men were dispatched to investigate the noise and soon came under intense enemy automatic weapons fire. In the initial burst of enemy fire both men were seriously wounded and became trapped between the assaulting enemy force and the team position. Simultaneously, Sergeant Wierenga and the remainder of his team received machine gun and automatic weapons fire from an estimated company size enemy force. As the Sergeant maneuvered to secure his radio and establish communications, he came under concentrated fire from a nearby enemy machine gun. Reacting quickly, he placed suppressive fire on the

enemy position and silenced the machine gun. He then rushed forward and threw several hand grenades at the nearest enemy positions. This action forced the enemy back and allowed the recovery of his two wounded comrades. Returning to his team, Sergeant Wierenga began to call in air strikes against the enemy. As the extraction helicopters arrived over the landing zone, he ran into the exposed area and directed the aircraft. Because of Sergeant Wierenga's determined efforts, the entire team was successfully evacuated. Staff Sergeant Wierenga's gallantry in action was in keeping with the highest traditions of the military service and reflects great credit upon himself, his unit and the United States Army."

1-1 SGT Steve Hoffman received the Bronze Star with a combat "V" device for his actions this day. His award citation reads as follows:

"For heroism in connection with ground operations against a hostile force in the Republic of Vietnam. Sergeant Hoffman distinguished himself by heroism on 26 April 1970 while serving as assistant team leader for a long range reconnaissance team operating deep within enemy territory. Shortly after leaving their night defensive position, the team heard noise approximately 60 meters to their northwest. Sergeant Hoffman immediately volunteered to investigate. Taking one of the indigenous commandos with him, he left the remainder of the team and cautiously advanced in the direction of the noise. After moving thirty meters, they spotted three North Vietnamese soldiers armed with AK-47 rifles. Closer observation revealed eight more well-armed soldiers. The enemy soldiers suddenly spotted Sergeant Hoffman and the commando and unleashed a *heavy volume of small arms fire on them. In the initial burst of fire the commando received several wounds in both legs and Sergeant Hoffman was wounded in his right thigh. Lying in an exposed position, Sergeant Hoffman sighted a ditch which would provide good cover. Although wounded himself, he crawled to the wounded commando and began to drag him towards the ditch, totally ignoring the grenades, heavy automatic weapons and small arms fire. Upon reaching the ditch, Sergeant Hoffman immediately began medical treatment on the commando's wounds. Then realizing his suppressive fire was not only required to protect his own position but was also to thwart the advance of the enemy against the rest of the team, he began to repeatedly expose himself to obtain better firing positions. When not directing fire against the enemy, he continued to treat the wounds of the commando. Through his professionalism and devotion to duty, he not only was the primary factor in repulsing a fierce enemy attack, but also saved the life of the wounded commando. Sergeant Hoffman's conspicuous gallantry and intrepidity in action were in keeping with the highest traditions of the military service and reflect great credit upon himself, Special Forces and the United States Army."*

RT ANACONDA's 1-2, SSG Donald Maley also received the Bronze Star with combat "V" device for his actions in this engagement. His award citation reads as follows:

"For heroism in connection with ground operations against a hostile force in the Republic of Vietnam. Staff Sergeant Maley distinguished himself by heroism on 26 April 1970 while serving as radio operator for a long range reconnaissance team operating deep within enemy territory. The team of three Americans and six

CIDG commandos was on an intelligence gathering mission. On the morning of the second day as the team began moving from their original position to relocate, they detected noises to the northwest. The assistant team leader and one of the commandos volunteered to conduct a quick reconnaissance of the area. After moving thirty meters they spotted eight well-armed enemy soldiers at a distance of only ten meters. The enemy spotted the two team members and unleashed a heavy volume of automatic weapons fire and grenades. Sergeant Maley observed the attack and saw the two men fall wounded. The assistant team leader was able to drag the commando to a covered position but they were still well within grenade range. Realizing the seriousness of their wounds and the helplessness of their situation, Sergeant Maley ran across thirty meters of exposed terrain to assist them. Upon reaching them, he discerned that he would not be able to help both of them at one time. He helped the assistant team leader back across the open terrain to the team's position, and again disregarding the heavy enemy fire, returned to the commando's side. Due to the severity of his wounds, the soldier was unable to walk, and Sergeant Maley was forced to carry him across the exposed terrain to the team's position. He successfully crossed the open area four times under heavy automatic weapons fire and engaged by enemy grenades, in order to rescue his wounded team members. Staff Sergeant Maley's conspicuous gallantry and intrepidity in action were in keeping with the highest traditions of the military service and reflect great credit upon himself, Special Forces and the United States Army."

CHAPTER 28:

THE PURPLE HEART. 11 AUGUST 1970

With the departure of the Cambodian team members from RT ANACONDA, it was time to rebuild. Instead of Cambodians, Dutch would be assigned a group of ethnic Chinese mercenaries, called Nungs in the local dialect. These fierce warriors were legendary for their courage and battle prowess. But Dutch would not be with them long after incorporating them into the Team. Dutch had a new 1-1 due to the medical evacuation of SGT Steven Hoffman, who had developed gangrene in his legs due to wounds received on the 26th of April. SSG Don Maley had likewise moved up to Team Leader (1-0) of RT ASP after the 26 April 1970 mission, the only patrol he ran with RT ANACONDA. This excellent soldier and new 1-1 was SSG Keith Kincaid. Dutch remembers Kinkaid as personally quirky, but also recalls him as a strong, capable and totally reliable 1-1, with a complete disregard for danger or difficulty. In time he took over as 1-0 of ANACONDA. On the 11th of August 1970, events took a turn for the worse.

DUTCH: *"So the Cambodians left, and I got a Chinese team. They were good! It was the second mission out with the Chinese team when I got hit. This was in August of 1970. We were going into an area in Laos, near the North Vietnamese border, that a team had been to three weeks before, but had been shot out of the LZ by heavy fire before they could even get on the ground. On that earlier mission, my team had been inserted at an LZ slightly earlier than the main team's insertion. We were hoping to draw interest, and we set up a defensive position with a bunch of explosives, claymores, LAWs, etc, so we could kill large numbers of whoever responded, while the main team quietly inserted somewhere else nearby for a reconnaissance patrol. Not long after we got on the ground and got everything set, here came the helicopters back! Huge numbers of enemies were out in the open at the main LZ, waiting for them and the fire was so intense, they had to abort. They told us to get back aboard, we were extracting. They had us blow everything we had set out in place. When we lifted off and our charges blew, there were huge secondary explosions; we were in a major ordnance/truck park.*

Three weeks later they asked me if I would go back again. I said, "yes, we'll try it out." So we studied it a different way, and tried coming in to the original planned LZ a different way. Well that didn't go too well. This time the enemy was pulled back into the woods nearby, waiting for us with lots of heavy weapons.

We flew out of Quang Tri, about an hour and a half at 9000 feet. We were on final approach on the Kingbee helicopter, and the helicopter was turning and descending. I was going in first. The Team Leader goes in first and comes out last. The Vietnamese had moved a lot of heavy weapons up on the hills above the LZ. Well, the first thing that got to me, it was like a Rambo movie. All the anti-aircraft weapons going off around you, next to the helo. The aircraft tail had over twenty holes in it, afterwards. I was sitting in the door over the ladder, facing the door gunner. He was just frozen in shock, not returning fire, until I hit him and told him to open fire. Then he responded.

Then I got hit. I didn't feel it. I didn't know I was hit. The ladder that we used was in the middle of the door. I was just thrown off it. I just tried to climb back on. The guy sitting across from me was my Nung Chinese medic. He was trained by Special Forces soldiers. So he told me "you're hit!" He showed me where I was wounded. (I had numerous 12.7mm heavy machine gun bullet fragments, a big gaping wound on my left leg thigh below the hip. You could see right through the middle of my leg from front to back, there were two big fragments behind and below the knee.) So I took my morphine syrette that we had, and sort of held it over the area, but hesitated to stick it in, because it hurt. He just sort of slapped my hand and it went in. We immediately aborted and returned to base at Quang Tri. I threw something at the pilot and got his attention, then hand signaled him to abort. I had told the other helo to abort by radio. Nobody else was wounded.

But that didn't bring an end to my Special Forces career. I went to the hospital. The Colonel pinned a Purple Heart on me in the hospital. They were going to send me to all these humongous hospitals back in the states. But I said "No, I'll stay in our little hospital right here." The Doctor told me that if I wanted to go Stateside, I'd be on the way the next day. But I stayed. The Doctors were looking at my wound like it was an old bicycle. I asked them what they were thinking. They said, "well, we're deciding whether or not to take off the leg. We're not sure we can fix this." I told them "if you take it off, I'll shoot you." Of course, I wouldn't do that. So they put my leg back together.

It was doing OK. I worked my way back, PT'ing on the beach. It took about three months to heal. When they made me leave the hospital, I moved into the senior medic's hooch and continued to recuperate. I had a total of seventeen 12.7mm bullet fragments in my leg. Two have worked their way out over the years, Fifteen are still in there. After about three months, they let me go back on missions. First it was short, one-day missions to implant sensors near trails and roads. We would insert with a team, implant the PSITS and then return the next day. So I took the Team back over after a while. I started out on small missions at first, short-range, short duration missions emplacing traffic sensors by trails and roads so we could monitor traffic from far away, things like that, until I got back on full duty again.

Then just before my first bigger mission, they called me in and told me that there was a medical issue with my wife. She was in the hospital so I had to go home. I went home in January of 1971, three months before the end of my tour. My planned DEROS date was in March of 1971. I was getting ready to go on a mission, and the Sergeant Major called me in and told me I had to go home right then. I packed my bag, and went down to the airfield, and boarded a C-141. They let me sit in the cockpit all the way to Okinawa. It was much better than being in the cargo hold. I took a commercial flight from Okinawa, and they put me in charge of all the other soldiers on the flight. That was no fun. They were asking the stewardesses for lots of alcohol, but I told them they could have only one drink apiece. They didn't like that. We flew into Travis AFB, and when we embarked in San Francisco airport, I and six to seven other SF guys were spit on in the airport by protesters. It actually happened. The airline put us in first class, and had us change clothes while they cleaned our uniforms and gave them back to us. They were nice about it, our uniforms were really clean. You forget about it. It was just one of those things. It was a night flight and I slept all the way back to North Carolina.

I got my wife out of the hospital within a couple of days, and we found and bought another house, as we had sold the first one. We lived there a long time, almost until I retired. When I came back, I asked Mrs A (Mrs. Billye Alexander, the Pentagon administrator who was the Special Forces Assignments Officer) to send me back to CCN to finish my tour, but she told me no, and got me an assignment to the RECON School at Ft Bragg."

CHAPTER 29:

SF RECON SCHOOL. JANUARY 1970 – APRIL 1974

Dutch was assigned to the Special Forces School at Fort Bragg, NC in January 1970. He was one of the most experienced reconnaissance team leaders in the US Army, and as a veteran of MACV/SOG, had instant credibility and exceptional expertise. He began instructing other Green Berets in the skills of recon patrolling which he had just months earlier been executing in the jungles of Laos, Cambodia and North Vietnam. His knowledge comprised literally a thousand small details of operations planning, use of terrain, movement, observation, communications, equipment selection and carriage, munitions selection, team organization and formation, and a myriad of other martial skills honed in direct combat in Southeast Asia. He passed along this priceless knowledge to hundreds of trainees, many of them among America's finest warriors who were themselves headed for combat in Vietnam. But the war was in its final stages, and combat operations there were winding down. While he was teaching at the Recon School, MACV/SOG was disestablished in May of 1972, and American combat involvement in the Vietnam War officially ended in January 1973. Dutch continued to instruct at the Recon

School until April 1974, passing along his hard-earned, priceless knowledge of reconnaissance patrolling.

While instructing at the SF School, Dutch had also encountered an old friend – and changed the course of his life.

"My old friend and teammate Steve Hoffman (RT ANACONDA's 1-1 in 1970) had been medevac'ed to the United States, as he developed gangrene in his leg, where he was wounded above the knee in the April 1970 mission firefight in Vietnam. He got out of the Army. I later met up with him in Fayetteville, and after our conversation, a week later, he was back in SF. I was not too popular with his wife after that. He had a great career – thirty-two years. He and his wife stayed together, they had two girls.

Late in his career, he was a Squadron Sergeant Major in Delta Force. He came to my house one day in late 1979, trying to get me to sign a piece of paper to get into Delta. There was something big in the works, and he wanted me. He said "C'mon, let's go!" I wouldn't sign it unless he told me what it was for, and he couldn't tell me. It turned out to be Operation Eagle Claw, the disastrous aborted rescue of the Iranian hostages on 24 April 1980. The next day, I was driving to PT in the morning, and heard about it on the radio. I called his wife, who told me that they were on their way back, and that most of them were fine. I ultimately learned that eight men had been killed in an aircraft collision on the ground. In my opinion, Delta Commander COL Charlie Beckwith lied to President Carter. He told him they were ready – they weren't ready."

STEVE HOFFMAN REMEMBERS THE ENCOUNTER WELL: *"In early March of 1971, I was working at IBM, and I was just so frustrated. It was not what I wanted. I got into contact with Dutch and my wife Debbie and I drove down there and stayed at Dutch's house for about five days. In those five days, everyone that was at Ft. Bragg that had been at CCN when I was* *there, Dutch invited over to the house....and that was it. I was just done. I was done with IBM. What happened was that they told me that I couldn't jump anymore due to my leg wounds, which meant that I couldn't be in SF. Well, if I couldn't be in SF, I didn't want to be in the Army. So, to get back in, I had to take a physical. I worked HARD to make sure that I could pass that physical, with my right leg. Work, work, work, all the time. When it came time for the physical, you know how they stand you up there and the doctor looks at you, "are you ok, etc."? He looked down at my leg. It's about a ten-inch scar. He said "what was that?" I said, "well, I got wounded in 'Nam...not bad." He said "well, it looks kind of bad to me!" I said "well, it doesn't bother me!" He asked "it doesn't bother you? Well, OK!" And I was back in Special Forces.*

I did try to get Dutch into Delta in 1979. He was the perfect guy for that."

While Dutch was an instructor at the SF School, he also had an interesting encounter with his family history. **HE REMEMBERS:** "It's kinda funny. I was an instructor in SF. In one class, I had two Indonesian students. The Army said "well, you speak Indonesian, you take these two guys." So we became friends, and began talking about Kalibening, which means "Clear River." We had had five swimming pools at the resort which my family owned when I was a kid. The property had been taken over by the Indonesian Army and made into their Officer Candidates' School. One of the students said "I'm the commander of that school." We laughed about it. I said "pay me back some money!" I never really thought about it. It was what it was. So he invited me to come and jump (parachute) into the school. But it never came about, it was a political thing. And then I retired, and came to work for CIA. Before reporting in, I had a break (vacation) with him. By then he was a bodyguard for the President of Indonesia, he had led the commando force that raided a nearby island and brought it under Indonesian rule. So as a CIA officer, I had to cut ties with him. He was a nice guy – he was a Catholic! His Deputy was a

Muslim. But they got together, got along. That guy was a hell of a shot. I remember the SF training course took him out to the firing range and the Range Officer showed him a gun. "Do you know how to use this?" He said, "yes, I know how to use that gun." I just laughed and stood in the back. And he just went through and knocked all the targets down. The Range Officer said "well, OK!" He did pretty damn good. He had been an instructor in the Indonesian Army."

Dutch was now 38 years old, and the war in Vietnam had ended. His assignment as a Recon Instructor at the Special Forces School had identified and honed one of Dutch's greatest gifts – that of being a skilled and dedicated trainer of younger, less experienced soldiers. Dutch in turn knew from his hard experience the value of sound training, delivered by knowledgeable, capable instructors. Being a skilled combat soldier does not automatically make someone a good teacher. That is a completely separate skill set, requiring patience, empathy, good communications skills and an ability to relate to a wide spectrum of trainees. Not many expert soldiers have this ability. Dutch had it in spades. So he was next assigned as an Instructor at one of the most demanding training programs in the entire US Military – the US Army's Free Fall Parachuting School, also located at Ft. Bragg, NC.

CHAPTER 30:

HALO SCHOOL.
MAY 1974 – AUGUST 1977

Most civilians think of parachuting as "skydiving" – an activity that is conducted strictly for thrills. But military parachuting is serious business. It is all about placing a group of soldiers on the ground, at a specific spot near a mission objective, close together so that they can form into their assigned units and proceed on mission tasking, with the equipment and weapons that they will need to perform that mission. By definition, these soldiers will be BEHIND enemy lines, the point of parachuting them in being to place them somewhere close to key objectives without them having to fight through an enemy's forward defenses to get there. The vast majority of military parachutists conduct Static Line jumps. The jump aircraft approaches the designated drop zone at relatively low altitude – usually between 500 and 1000 feet above ground level if it is a combat jump. Each soldier's round parachute is connected by a static line to the aircraft, so when the soldier jumps, the parachute is pulled out of its pack tray and automatically opens. The soldier has limited control of the parachute's flight direction and he reaches the ground fairly quickly, usually within a few seconds of jumping. The parachute is designed to descend at about sixteen miles per hour, a compromise speed intended to get the soldier and his heavy personal equipment load to the ground safely with minimal chance of injury (but landing casualties do occur, not infrequently.) A slower descent speed would leave him hanging in the air over enemy territory for a longer time, vulnerable to increased wind drift and ground fire. Obviously, a 220lb soldier carrying a mortar tube and ammunition is going to descend faster than a 120lb soldier carrying a correspondingly lighter personal load. The soldier hits the ground fairly hard, performing a "Parachute Landing Fall" (PLF) in an effort to prevent injury, and removes his parachute. He turns into a light infantryman as soon as he does. The parachute is abandoned, its job complete.

A much smaller subset of military parachutists conduct Free-Fall parachute operations. These are far more technically-demanding jumps, requiring a much higher degree of training and expertise to complete safely. Given that the mission almost certainly requires secrecy, the small teams of jumpers do not wish to be detected, so combat jumps most likely will take place at night, with each soldier wearing Night Vision Goggles (NVGs.) The jump aircraft is at a vastly higher altitude than for a static line jump, usually 15,000-35,000 feet above ground level. Above 22,000 feet, there is insufficient oxygen to sustain life, so the soldier will be breathing supplemental oxygen. It is also very cold, perhaps as much as -45F, so extra clothing must be worn to prevent frostbite. The aircraft is so high that it is far less vulnerable to enemy anti-aircraft systems, and it may in fact be flying in airspace outside the national borders of the country where the drop zone is located. The parachutes used are rectangular in shape, and are steerable ram-air canopies with wing-like characteristics and a significant glide slope. They can fly long distances across the ground, steer precisely and flare before landing, slowing the descent rate even further and allowing a gentle, stand-up landing at an exact landing point. When it is time to jump, two different techniques may be used.

High Altitude High Opening (HAHO) parachutists exit the aircraft, get stable in free fall, and almost immediately open their chutes. They then remain close together and navigating with altimeter, compass and GPS (or to a pre-positioned infrared beacon on the drop zone seen through their NVGs) they fly as much as thirty or more miles across the ground to their precise intended landing zone. This process takes as much as half an hour or more under canopy.

High Altitude Low Opening (HALO) is the most demanding technique of all. The jumpers exit the aircraft, get their bodies stable in flight, and optimally, they physically link up, holding on to one another so that they remain in contact. They free fall towards the earth at a terminal velocity of 120-200 miles per hour, depending upon body position. While in free fall, they present very little radar return and are extremely difficult to detect. At approximately 3000 feet, they track away from each other and open their parachutes, attempting to remain close enough together to allow joining up easily when they land. They navigate to their desired Landing Zone and after landing, link up and begin their mission.

The first experiments with high altitude parachuting took place in the 1950s, and the HALO technique was first used in combat on the 28th of November, 1970. The six-man MACV/SOG Recon Team FLORIDA jumped into Laos from 18,000 feet altitude. The American members of that team were 1-0 Staff Sergeant Cliff Newman, 1-1 Sergeant First Class Sammy Hernandez, and 1-2 Sergeant First Class Melvin Hill. The others were two Montagnards and a South Vietnamese Army officer.

The techniques and procedures of HALO Parachuting were subsequently further developed and refined at the US Army's HALO school at Ft. Bragg, NC. The Army intended for this groundbreaking technique to continually evolve, always taking advantage of new materials, equipment designs, instructional techniques and better tactics as time progressed. To that end, some members of the HALO School instructor cadre were tasked to form the HALO Committee, the brain trust of this most highly-demanding and technical military art. This Committee of the most experienced experts in the art of military free fall parachuting was and is responsible for recommending changes to the approved procedures and equipment when advances occur. And Dutch was soon to become a charter member of the very first HALO Committee.

DUTCH RECALLS: *"In 1974, I went to the HALO school, and after graduating, they kept me in HALO school as an Instructor. I had already logged a lot of free fall jumps when I got there. SGM Morecon told me "you've got to go through as a student first, to get your Free Fall License to be an Instructor." They gave me all kinds of hell, as "extra instruction." I became a member of the first HALO Committee, a group of about 12 free-fall experts, and I did that for just over three years. We ultimately were all E-8s. This was 1974-1977. The school graduated hundreds of students during that period. We taught Basic HALO, HALO Jumpmaster and HALO Instructor courses. The basic HALO school was four weeks long. There would be about thirty to thirty-five students in each class, and usually only about four or five would not graduate – they just couldn't do it (unless they were a Colonel, in which case they graduated anyway (chuckling.))*

Each instructor was given responsibility for two to four students to personally work with on all their training jumps, so they always had a familiar person flying with and correcting them in free fall. It always seemed like I always ended up with the heaviest guys, who were the most difficult to stay with and adjust in flight. All the students were qualified static line parachutists, but almost none of them had ever done any free fall before. The first two weeks was ground school. We would have them lie on table tops, and teach them the various body positions to control their descent rate, direction, and stability. We taught them how to rig their equipment so it didn't come loose and cause instability or injury to them. I really enjoyed it – it was great. I got to jump for free!"

CHAPTER 31:

ROTC AND SPECIAL FORCES TRAINING. SERE SCHOOL. RETIREMENT.

In 1977, Dutch was finishing up his assignment at the HALO School and Committee. He had been instrumental in training hundreds of US servicemen in the techniques of free-fall parachuting, and had made thousands of jumps, training students, helping to refine these techniques, testing new parachute and equipment designs, and recommending improved procedures for adoption. It was time to move on to another assignment. Another area of Army training which needs good instructors is the many Reserve Officer Training Corps units at dozens of US universities. As a highly qualified SF soldier and combat veteran, Dutch would be able to inspire and instruct future Army officers, teaching them basic soldier skills while role modeling officer/NCO cooperation and professional relationships. It would also be an assignment offering a break from the extreme physical and mission demands of the HALO School. Dutch would be able to decompress a bit, spend more time with his family, and live in a far less regimented, purely civilian society in a northern city while helping build future Army leaders.

"I left the HALO School in 1977, and I was assigned to the ROTC at the University of Vermont. I was ready to go somewhere else. I went to ROTC training. I was supposed to go to a SF team in Germany, but for some reason one of the other senior NCOs was sent there and I was sent to be an ROTC instructor in Vermont. It worked out pretty well, I enjoyed it, it was good training. I gave all the tactical training, teaching rappelling, things like that at the University of Vermont Army ROTC unit for two years. That was in September of 1977 – September of 1979."

After two years away from the mainstream of Army life, Dutch was ready to go home and back to work supporting the active duty army. He received a well-deserved promotion to Master Sergeant (E-8), and was now expected to take on a significant senior enlisted leadership role at a large combatant unit. But when offered the chance, Dutch was very happy to take on an even more important challenge; that of training the next generation of Green Berets. He was assigned as the NCOIC and Senior Enlisted Instructor at the Special Forces Training School at Camp Mackall, Ft. Bragg, NC. He was returning as First Sergeant to the School where he had graduated and earned his Green Beret ten years earlier, in 1969. As always, Dutch worked extremely hard, led by example, and set high standards for himself, his Instructor cadre and his students.

"They wanted me to go to either the 82nd or 101st as a First Sergeant because I had just made E-8, but I was lucky and was allowed to go to Camp Mackall as First Sergeant. I never, ever expected to rise that far in the Army. I didn't want to go to a "Leg" unit, and they asked me if I wanted to go to the Special Forces Training Center, at Camp Mackall.

So from September 1979 – May of 1980, I had my tour as First Sergeant at the training facility, Camp Mackall, which is now a hotel. It wasn't a hotel then. We walked the buildings over from another location to where it is now, another part of Camp Mackall. It's really doing well now, it's really big. It was okay. I made a lot of changes, changed a lot of the instructors, and their responsibilities. I went out on runs and rucksack marches with the students; the previous First Sergeant hadn't. The original Seal Team SIX came through the Recon School under me. I put the spurs to them. I took a big Navy knife away from one of the SEALS who was carrying it without authorization. I told him "if you think you are man enough, you can come and get it from me at the office." He never did. I had it a long time, and I gave it to my grandson.

Classes were between 70-90 men each class. About 10-15 would immediately quit upon arrival. We would ask them "any of you that don't want to be here, or don't think you belong here, go ahead and step back now. After this, there is no stepping back." About ten to fifteen would get back on the bus and go back to their old units. Then in training, another 10 or fifteen would either quit or be injured, and unable to finish. So about sixty of ninety men, or about two thirds, would successfully complete the course. I visited again in later years, and watched a busload of students arrive. TWENTY FIVE of them quit upon arrival. I couldn't believe that."

In due time, Dutch wanted to go back to lead an active SF team, and the Army wanted to put all his skills and leadership abilities to that important use. He received orders to Bravo Company, Second Battalion, 7th Special Forces Group, as a Team Sergeant.

"When I came back to the active SF in June 1980 after my tour as First Sergeant at Camp Mackall, I took over a team in the Second Battalion, 7th Special Forces Group and trained with them for a couple of years from June 1980 – January 1982. I had a pretty good team. We were based at Smoke Bomb Hill, on Ft. Bragg. It used to be barracks – now there are buildings there. During that time we completed three team ARTEPs, two actual missions and a thirteen-month Mobile Training Team deployment to Monrovia, Liberia, to train the Liberian Light Infantry Battalion, their so-

called "special forces." One memorable event was the installation of some radio equipment on top of a hundred-foot tall tower. They wanted a radio up there, but had made no provision for climbing it or getting the equipment up there. Finally, there was no other way, so I had to climb up there and pull up the equipment and install it myself.

I really enjoyed it. Leading this team was a culmination experience – it allowed me to use all the training and experience I had gained over my whole career. I initially had a good team commander – CPT Miller. He had been an NCO before that. He listened. Our wives were friends, and spent time together while we were away. Later, when we deployed to Liberia, I had a new First Lieutenant, an Academy graduate. He was about 6'4", maybe 6'5", and all he could think about was eating. We were in Liberia for a total of thirteen months – out of that whole time, he spent maybe one month outside his room. He wouldn't come out of his room, saying "It's too hot." I had one older Hawaiian guy – he was a drunk, I finally kicked him out of the team. We were in Alaska at that time. He had a few drinks one night, and the next morning, didn't make formation. I found him asleep, in his bunk. I told him to get up, he wouldn't. So after warning him one more time, I just dumped him out of his rack, upside down, covers and all. I told him "when you get dressed, report to the Sergeant Major – you're fired. You're off the team." He said "the Sergeant Major will believe me, you'll see!" He never came back. I had another guy – he was senior enough to be promoted to E-6, but I couldn't promote him, because he couldn't balance a checkbook. I taught him how to do that; guess what? He got promoted to E-6. He ended up as a W-3! He was a hard-working guy. I had two team members make Sergeant Major."

It is clear that Dutch took his leadership responsibilities seriously and performed them capably, leading his team by example, and helping each man reach his full professional potential while accomplishing their assigned missions. His leadership talents honed the skills of others, giving the Army several more future senior enlisted leaders to pass along these abilities to new generations of soldiers.

"It was kind of a hard time. I kicked the Lieutenant out – he was no good. I had enough status and authority that when I told the Command that he needed to go, they just trusted me and listened to me. He later encountered my mother when we were attending an air show as a family at Pope AFB. She was visiting me at the time. We were all standing around a plane, and he happened to be there. He told her "your son was hard on me!" My mother wasn't sympathetic."

In 1982, Dutch felt that he was nearing retirement age, and that he had risen further in the Army than he had ever expected to. He thought it might be time to close out his Army career. But a legendary Special Forces leader and Vietnam War hero had other ideas. Dutch was asked to help establish what is now a cornerstone of American military combat training– the Survival, Evasion, Resistance and Escape (SERE) school.

This important training school was established by SF COL James "Nick" Rowe, and its demanding curriculum was largely based upon his personal experiences of having been held as a POW by the North Vietnamese for five years in extremely harsh jungle conditions. He eventually escaped his captivity and returned to the USA. He later wrote an informative book, Five Years To Freedom about his experience. It remains a classic of professional military literature.

"I was going to retire at just over twenty years of service in 1983. But Colonel Nick Rowe had returned to SF, and had found out about my childhood POW experience. He told me "no, you're not. You're going to help me set up this POW training school." (This is what we now know as SERE: Survival, Evasion, Resistance and Escape training, and many thousands of servicemen have attended it.) Well, I got a team together, all young NCOs and we

put the school together from February 1982 – January 1984. I was the NCOIC and Senior Instructor. The only thing we didn't have at that time was the RTC, where they have the interrogation training. This was also at Camp Mackall.

COL Rowe and I used to spend our lunch periods together. We would take a hike of about three miles away from the Training Center, and then sit and eat a bag lunch, talking. When we were finished, he'd call a car to drive us back. Other times, he loved to go snake hunting. He'd have me check two shotguns out of the armory, and we'd go wading in Big Muddy Creek, also called the "Dead Man's Creek," which ran through Camp Mackall. It was named that because supposedly slave owners would drown runaway slaves they wanted to get rid of in it. We'd go wading in the creek, shooting water moccasins until we had a bunch of dead snakes. Then we'd go back."

After the SERE Program was up and running smoothly, the JFK Special Forces School turned to Dutch to establish another landmark specialized training program. He was assigned to this final task from February 1984 – December 1985. It would be his final contribution to the US Army's training mission. But the Army itself was changing.

"I started to set up another school for Long-Range Reconnaissance Underground – missions where you go in early, at night, and you have enough time to dig the holes and set everything up and stay underground for up to 21 days. Well, with one General, the Commander of SF, I made a kind of big boo-boo there. He told me, "our guys aren't going to be willing to stay underground for 21 days. They're not Chinese, not Vietnamese, they're Americans. I know how much money you make, and you know how much I make. I'll bet my pay against yours that they'll never make it." So we settled on 14 days. The first time, we got a call after a few days that someone was out of their hide, on top. Well it was the Team Leader, an officer. Then they told me that the 4th of July

was coming, and they wanted me to either bring in the team, or take special food out to them to celebrate it. I told him, "Sir, the war doesn't stop" – meaning that I didn't want to do that. It was contrary to the focused realism of the training experience that we were trying to impart. I refused to do it, and that didn't go over well. So, that's when I left. I left the school to them and started another training program, similar to LRRSU. It was called the Advanced Land Reconnaissance Course (ALRC.) I understand that some of the elements of that course have been put into action in Afghanistan.

By the end of that assignment, I knew that it was time for me to go. I went back to COL Rowe, and told him that I had had enough, I was retiring. I told him why, and he said he could understand it, so I retired with over 23 years of service. I retired from the Army on 01 January 1986."

CHAPTER 32:

A NEW BEGINNING. THE CENTRAL INTELLIGENCE AGENCY. 1986-1999

Dutch's life and career in the service of the United States was not yet drawing to a close. He was far too active and far too young to even consider permanent retirement at that time. He was only fifty years old, and in good health and physical condition. He had not yet decided what to do after his retirement from the Army when the phone rang unexpectedly one day.

"I was in the midst of retirement, not really working anymore, and I got a call from a phone number in DC that I didn't know and they said that they wanted a shooter. Well, I told them that I didn't shoot anymore, I was really kind of tired of it at that time. But I decided to give them a call back. It was a bunch of old friends of mine that had kind of disappeared over the years. And they worked for the CIA. They encouraged me to apply, gave me the information I needed to submit an application. I filled

out all the forms, took several interviews, and successfully went through the polygraph examination. I made it through all of that fine, and they hired me as a GS-11. So I retired in December of 1985, and by February 1986, I was working for the Agency in Langley, Virginia. Again, I was the oldest guy going through initial training."

SSG (Later MAJ) Henry Kohn served in 5th SFG's Project SIGMA, and later in MACV/SOG CCS First Reaction Company and on several Spike Teams in 1968-1969. He saw combat several times defending small radio relay sites on Nui Ba Den and Nui Ba Ra mountains. He returned to Vietnam after the war ended, serving with the Joint Casualty Resolution Center (JCRC) trying to recover the remains of US MIAs and soldiers and airmen who were killed without their bodies having been recovered. He later joined CIA and served as a Paramilitary Officer and later as a Contractor for over thirty years. He retired from CIA as a ▓▓▓ (a civilian grade analogous to a military O-6 Colonel). Henry is a personal friend of both Dutch and the author, and was for a time our direct supervisor when we were serving together as Instructors at the CIA's training center, known as The Farm. He recalls: *"Dutch put me through RECON school in 1971, and through HALO School in 1974 and HALO Jumpmaster in 1975, all at Ft. Bragg. Dutch is the only one of my former SF friends who I sent an application to, and who then successfully made it all the way through the hiring and security clearance process and was hired at CIA. Dutch was universally liked and respected by everybody who worked with him."*

Chartered in 1947, the Central Intelligence Agency has several responsibilities in support of the Executive Branch of the United States Government. It is a large, sprawling bureaucracy, ▓▓▓▓▓▓▓▓▓▓▓▓▓▓▓▓▓▓▓▓▓▓▓▓▓▓▓▓▓▓▓▓▓▓▓▓. (lots of people of various types work there.) Clandestine intelligence collection from human sources, and covert action operations are conducted ▓▓▓▓▓▓▓▓▓▓▓▓▓▓▓▓▓▓▓▓▓▓▓▓▓▓▓▓▓▓▓▓▓▓▓▓▓▓. (by some of the people who work there.) In 1986, the CIA had four main directorates: the Directorates of Operations, Intelligence, Administration, and Science and Technology. The Directorate of Operations (DO) was the clandestine arm of the CIA, and within it, the Special Activities Division (SAD) was the paramilitary arm tasked with the conduct of covert action operations in response to Presidential Findings. SAD was where the Paramilitary Officers worked, putting their hard skills to use as directed by the President and National Security Council.

SAD ███████████████. (has several subordinate parts.) These included ████████████████████████████████ and the Special Operations Group (SOG), a direct descendant from the World War II Office of Strategic Services, the CIA's immediate forerunner agency. SOG was the Agency's paramilitary arm, ████ ████████████████████████████████ (organizational/structural details of the SOG sub-units). Dutch joined this small but elite cadre of professionals, of course being posted to SOG ████████. Among his first tasks was to learn how to access Agency data systems, to write and read intelligence reports, request name and other data traces, and to access as needed and authorized all ████ operational traffic ██. (CIA takes certain steps to protect classified information and the identities of its covert employees.)

"When I got hired, I was in a room with another old SF buddy, MAJ Jerry Poelking. Jerry was an exceptionally experienced HALO Instructor and tactical parachutist. I knew him from the HALO Committee. He was in one corner, and I was in another. They handed each of us a piece of paper, with the pay grade they offered to hire us at. He asked me "what did you get?" I told him "▮*" (a civilian grade analogous to an O-2 First Lieutenant) and he said "Mine only says* ▮*!" (a civilian grade analogous to an O-1 Second Lieutenant.) I said "scratch it out and write in* ▮*!" (the higher grade.) He did, and they gave it to him. We were assigned to Special Activities Division, SOG* ▮*, the paramilitary department of the Agency. The Chief of it then told me "You are going to computer school." I thought, 'my God, computers.' That was okay. I didn't learn all that much, but I did learn enough to get by with it."*

After getting settled into the Agency's headquarters bureaucracy and assigned a small desk in the SOG offices at Hqs, it was time for Dutch to go to work, supporting the daily missions of the CIA. Unsurprisingly, most of these activities do not take place behind a desk in the air-conditioned offices of Headquarters. They happen out in the field, in any one of the ▮ CIA Stations and Bases worldwide, or in covert training facilities maintained to support them as needed by mission requirements.

"Then I began the first of many temporary duty deployments overseas, called TDYs, with SOG. I did these kinds of trips for a while. My first real SOG ▮ *project was a training program. I was one of three SOG officers who were assigned to work on a covert action program focused on* ▮*, (a north African country) which was then a dictatorship run by* ▮ *(a known historical figure who died over a decade ago and who as of this writing is still dead.) He had been a thorn in America's side for quite a while. Our mission was to train a group of native-born, naturalized American citizens to be anti-*▮ *(the still-dead Dictator) guerrillas, and to teach them how to conduct*

infrastructure attacks. It was believed that these attacks would cause economic damage, and demonstrate that ▬ *(the still-dead Dictator) was not in control of his country. It would weaken him economically and politically, and might lead to his ultimate downfall. A major aspect of the plan was to use 90mm mortars to attack specific major petroleum pipelines. We trained the team of these individuals in a variety of combat skills – they got pretty good! We started with twenty-one, and ended up with twelve men when the training was over. These were good men, honest and true Americans, dedicated and wanting to do a good job to liberate their homeland. Of course, they still had their unique cultural quirks. For instance, when they needed to throw grenades, they always had to pray first. That was typical of Arab/ Muslim troops. Cultural or religious idiosyncrasies are always a huge issue when training or working with indigenous troops. You just have to recognize them, and work with or around them without getting frustrated. You can't turn foreign-born trainees into little copies of American soldiers, nor should you want to. They are effective in their area of operations specifically because they fit in there. This is something every Green Beret learns.*

This intensive training program took place at a couple of covert field training sites the Agency operates in the ▬ *USA, as well as one part which was conducted at a controlled-access area of a major* ▬ *(a branch of the US military) base. We took them to that base for small boat training in the ocean. I had taken a week off, but was called back unexpectedly. I asked "what's the problem?" Well, there was this one* ▬ *Major who had been assigned to support our training. He was extremely foul-mouthed and abusive. The* ▬ *guerillas just refused to work with him after a while. They literally refused to come out of their tents to do training with him around. We had him fired from the assignment with one phone call.*

Things immediately improved, and the ▬ *commandos went back to training, and again, they were doing really well.*

But cultural norms again reared their counterproductive heads. When it came time for their final exercise, the day before, we told them to cease showering, using strongly scented soaps, colognes, etc. This is routine bush fieldcraft. Well, they didn't comply. Being heavily perfumed is a cultural thing with a lot of Muslims. On the morning of the exercise, we checked them out, told the team leaders how we wanted them to move as a patrol. We took up an over-watch position along their patrol route. In a perfect world, they would have been able to silently sneak past us. Not this day. I told "Phil", my SOG senior partner on this program, "I think I can smell them." Nobody believed me at first, but I could; they were coming, right then and we detected them. (After that, we ceased using perfumed soaps and shampoos at our SOG field training sites' gym and bathroom facilities.) Then they got to enjoy a couple of days in a nearby coastal town, relaxing and doing touristy things to unwind. They brought me a hugely tacky souvenir of mood lights, to hang up in my living room. I just couldn't put it up, it was so garish. Fortunately, they never visited my home, so they weren't offended. We pronounced them ready to deploy; they were skilled enough, had completed the requisite training, and it was time to send them on their mission. We had them in isolation at one of our covert training facilities. And when it was time to make the decision to launch the mission, Headquarters got cold feet and canceled it."

This is not unusual in the risk-averse Agency bureaucracy since the 1980s. What used to be a culture of "we can do it – mission first" in the CIA has gradually morphed under uninspired or incompetent leadership, hostile Congressional Oversight, a relentlessly oppressive climate of political correctness and other media-driven pressure into a pale shadow of its former self, afraid to do anything which might possibly fail or cause political trouble. What is the process of paralysis? The President signs a Finding authorizing a covert action in support of specific US policy goals. Plans are conceived, refined and proposed to carry out those actions, and the best plans are approved up a chain of command.

Bureaucrats authorize funds and give permission for mission preps to take place. Mountains are moved. Huge sums of money are spent. Supplies and equipment, some of it extremely sensitive and specialized, are acquired. Good men work tirelessly over long hours both as trainers and trainees. They endure physical hardships and family separations and they take significant personal risks to prepare and train for their assigned mission. They succeed and get themselves able to do what they have been tasked to do. They are ready. All is prepared.

And all this time, weak-willed risk-averse middle and senior managers (most of whom have exactly ZERO life experience of this type, and therefore can't conceive of what the team is capable of doing, has been made to do and has willingly done) are silently watching and waiting. These so-called Managers sit behind their desks in the comfort of their air-conditioned offices in Langley, Virginia or in the National Security Council or Congress in Washington, DC, hoping that somewhere along the way, something will go wrong or conditions will change such that the mission will become impossible or so moot as to be unnecessary. That way, they won't have to be responsible for running the risks it entails, or for making a specific decision to either execute it or cancel it. Sometimes, circumstances do intervene and result in mission cancellation with no one having to make a hard decision, thereby extending top cover for their bureaucratic cowardice. But all too often, when it is time for the rubber to meet the road, these "Executives" get cold feet, and push the "abort" button at the last second, rather than taking responsibility and incurring the risks that they previously claimed to believe were warranted in pursuit of US interests. And all that came before is instantly rendered for naught. Is it any wonder that most Paramilitary Officers loathe Headquarters bureaucrats?

On this occasion, at the eleventh hour on the eve of deployment, Headquarters canceled the ▓▓▓▓▓ covert action mission. Someone decided that their career interests would not be well-served if these men went about causing loud noises and disturbing events inside ▓▓▓▓ (the still-dead Dictator's country.) It is far easier to say "no" when risks are involved....and thereby keep moving smoothly up the promotion ladder. Now in light of his long career in SF, this was not Dutch's first covert action rodeo, and he knew that the next step in the slow dance of bureaucratic death would be to hold the team on site, "awaiting a better time to execute." This could extend for weeks, and ultimately the whole program would still be scrubbed – it would just look like this was a result of a deliberative process, rather than a knee-jerk act of managerial cowardice. The same risk aversion would ultimately deal the death blow to the program; it would just be parceled out over a drawn-out period rather than as a single, discrete decision. Again, the desk weasels could spread the blame or responsibility around and dilute it to minimize their personal liability. So the Team arranged a covert aircraft to transport the trainees back to where they came from, getting them back to their homes and off CIA's facilities.

DUTCH RECALLS: *"I knew that Headquarters was dithering. I didn't care. So as quickly as possible, I told the pilots to load them up and take off en route to their destination. And they did. Sure enough, not long after takeoff, the phone rang. It was Headquarters calling to tell us to hold the team on site"to await a better time." I politely informed the caller that it was too late; given that the mission had been canceled, we had disbanded the team and sent the men home; the plane was already in the air. Years later, I saw one of them in a little bar in* ▮▮▮▮▮▮ *, Virginia. He called out to me, but I acted like I didn't hear him and walked on... there was nothing left to say. These were good men, loyal Americans, who wanted to do the job."*

And the files on this project joined countless other stillborn covert actions gathering figurative dust on electronic shelves deep in the bowels of Headquarters. Some faceless bureaucrat breathed a silent sigh of relief and went back to his or her resume' polishing.

Before long, Dutch was given another peculiar assignment, but of relatively short duration.

"For a time I was also assigned to a special project: babysitting a ▉▉▉▉▉ *(Latin American) pilot who was being kept in isolation at a military base in Maryland. For some reason, he had to be supervised and no one was allowed to see him; not even his wife. I don't know why. That was boring and slow. Finally, he became very difficult to deal with.*

Then I was asked to go to the Farm (The CIAs' legendary main operations training base) *as an instructor on a temporary basis for a while. I did that for a couple of months, working on a Free-Fall parachuting course for some Paramilitary Officers among other things. When that was over, the Chief of* ▉▉▉▉▉ *(my part of SOG), Jim Monroe, told me he wanted me to go back to The Farm on a* ▉▉▉▉▉ *tour as an Instructor. I told him I didn't want to do that, but he said "You're going – because so am I and I want you there."* So Dutch became a Farm Instructor for an ▉▉▉▉▉ assignment, 1986-1988, teaching CIA officers a variety of paramilitary skills. He would return to the Farm to teach temporarily a few times over the next several years. Dutch finally went back for a permanent assignment to the Farm in 1996, and he would remain there for over twenty-five years, teaching. This type of training will be covered in detail in later chapters. But before that, Dutch would be an integral part of one of CIA's most well-publicized and controversial covert action programs in its long history – support for the Nicaraguan "Contras."

CHAPTER 33:

THE CONTRAS. 1988-1991

Then a major chapter of Dutch's Agency career opened. He was sent to ▓▓▓▓ (a Central American country), first on a temporary basis and later in a Permanent Change of Station (PCS) assignment, to train and work with the anti-Sandinista Nicaraguan insurgents, known as the Contras. The Sandinista Government in Nicaragua was a hard-core Marxist regime, which had come to power in 1979 and was advancing the cause of communism in Central America in ways that undermined regional stability and damaged US interests. The US government therefore viewed the Sandinistas as a threat to the economic interests of American corporations in Nicaragua and to US national security. So early in 1982, President Ronald Reagan signed a top-secret covert action finding, National Security Decision Directive 17 (NSDD-17), giving the CIA the authority to recruit and support the Contras. It originally allotted $19 million worth of military aid. The effort to support the Contras was one component of the Reagan Doctrine, which called for providing military support to movements opposing Soviet-supported, communist governments.

"In total I spent over three years working with the Contras. I basically trained them in all aspects of insurgency and combat operations. This included teaching the Contras to shoot the "new" IR guided anti-aircraft missile – like an improved "Stinger" man-portable air defense system (MANPAD.) I was assigned to this task with "Phil", a more-senior SOG officer ▮▮▮▮▮▮▮▮▮▮▮▮▮▮▮▮▮▮▮▮▮▮▮▮▮▮▮ ("Phil" is now retired and his CIA employment cover has been lifted. Some of his other operational history post-September 11th 2001 is available on the open record. I refer the reader to Gary Shroen's excellent book **"First In: An Insider's Account of How the CIA Spearheaded the War on Terror in Afghanistan".**)

Phil had earlier been my partner in the ▮▮▮▮ *training program* (targeting the still-dead Dictator.) *We had to come up with some kind of training apparatus to teach the Contras how to track a moving thermal target with the new missile's seeker unit. Phil first tried a high-tech solution: he went out and using Agency funds, he bought a model aircraft to use as a target trainer. We found a field in Northern VA to test it, and launched it – he immediately crashed it. Turns out, flying a model airplane is harder than it looks! Then he suggested that we should buy a model helicopter. I gently told him that flying a helicopter actually took real skill! We nixed that idea. Ultimately, we put a lit cigarette on a line stretched between trees, where we could pull it along with a small string, teaching them to track it using the thermal sight mechanism. That incredibly low-tech solution served us well in the jungle training camps in* ▮▮▮▮▮▮▮. (Central America.) *Looking back, we were way out on a limb in those early days. Our international travel was in alias, of course, and our "*▮▮▮▮▮▮▮▮▮▮▮▮*" (fake story to explain our travel) had very little to no backstopping and zero preparation. We* ▮▮▮▮▮▮▮▮▮▮▮▮▮▮▮▮▮▮ (did nothing to substantiate the story.) *If we had ever been arrested, it would have fallen apart immediately. We spent about a year, working with different groups of Contra forces, in several covert training camps and marshaling areas inside* ▮▮▮▮, *(a nearby Central American country) advising, training and supporting them as they ran cross-border insertions into Nicaragua to engage Sandinista government forces and targets."*

One interesting side note: there was a popular piece of improvised jewelry worn by many of the Contra soldiers, called the "Contra cross." This was a 5.56mm cartridge, emptied of powder, but with the bullet still crimped in the casing neck, and with the back half of the cartridge case pressed flat and filigreed into a simple cross. It was worn around the neck on a stainless-steel bead chain. If you ever see one of these, it likely marks the wearer as having been associated with the Contras.

In time, and in recognition of his professional and leadership abilities, Dutch was moved up the ladder into a more senior leadership role. He was made the equivalent of a Battalion Commander.

"Then I got a three-year Permanent Change of Station (PCS) assignment; From 1989 through most of 1991, I was officially posted to ▮▮▮▮▮▮▮▮▮▮▮▮▮▮▮▮▮▮, *(the nearby Central American country) and took over as the primary CIA adviser to one of the battalions of Contra guerrillas, in fact acting as their Battalion Commander. I provided them with their intelligence support, saw to their health and welfare, paid them, supplied them, kept them organized and disciplined and sent them on their missions. We trained them on weapons, tactics, infantry skills, as well as on how to use systems like what we called "Trojan Horse." This was a clever insertion method, using a false compartment in a freight carrying cargo truck. They would cross the Nicaraguan border at nighttime, with team members concealed in the compartment. Even if the truck was searched, they wouldn't find anything. It worked every dang time, getting Contra soldiers into and out of Nicaragua for operations. It was a good assignment, and we supplied them well. They had*

more ammunition available to them than the US 82nd Airborne troops did, more explosives. They got more live fire training than our own soldiers did at that time. They were very well equipped and well trained. The Contras did fairly well in combat against the Sandinista forces. They were dedicated troops, and conducted numerous successful missions. They conducted every engagement, every mission they went into Nicaragua on according to plan, except once; one of their helicopters was shot down, and crashed in ▇▇▇▇▇▇▇. (an adjacent Central American country.) *At first everyone was panicking, thinking that this was going to cause a major international incident, inflame the* ▇▇▇▇▇▇ *(Central American country's government) and create great embarrassment for the Reagan administration. I counseled calm as we sorted out what actually happened. In fact, the helicopter had been downed by local criminal elements, not the Sandinistas or the* ▇▇▇▇▇▇▇ *(Central American country's) military. There was no fallout whatsoever.*

It was going pretty good, but after three years, it was time to go home. US policy changed, and support to the Contras was dwindling. At first, the Agency wanted me to be ambiguous, to keep the Contras in the dark about the future of the program. I strung them along for a while, but ultimately, I told my superiors that I had to tell them something. I was given the approval to inform them that the program was ending. I was forced to defund the Battalion, cut their wages, separate men and recover or destroy their weapons and explosives. This made me very unpopular with many of the lower-ranked troops. When the Battalion leadership learned that I was in some danger, they posted members of the best unit of the Battalion, the Reconnaissance Platoon, around my quarters to guard me. One of the Companies at first refused to surrender their weapons. They were simply not going to turn in their guns. I told them that there were two other battalions in the immediate area, and that if necessary, I would order them in to collect their weapons from them by force. They grumbled, but they knew I meant business and that they would inevitably lose that fight. They turned in their guns. When all was said and done, I paid all of the Battalion members off, we had a final assembly and parade, and made a few speeches. And with that, we disbanded the Battalion and sent them home. I departed ▮▮▮▮▮▮▮▮ (the Central American country where I was assigned) *in 1991.*

In 1990, the Sandinista government lost a popular election and fell from power. The people of Nicaragua had had enough of the war, enough of the inevitable fruits of Marxism which are oppression, poverty and despair. The Sandinista government had been proven to be ineffective and was summarily replaced. Nicaragua was returned to non-communist control. And the Contras had played a major part in bringing all that about. So despite all the later angst and hand-wringing about supposed misdeeds by the Contras, the fact is that they won the civil war against the Sandinistas, and positive change in both the USA's and Nicaragua's national interests was effected by the program.

CHAPTER 34:

WEST AFRICA. 1994-1996

CIA Paramilitary Officers are first and foremost Operations Officers; in addition to their paramilitary skills, they are expected to be able to spot, assess, develop, recruit, train, handle and terminate human agents, thereby collecting and disseminating valuable clandestine intelligence reporting. This is what traditional Operations Officers do. But some environments simply require a higher degree of tenacity or survival skills for an officer to perform well in them. Such austere and dangerous environments are the natural venues for Paramilitary Officers to be assigned to on a long-term basis. It is not unusual for a Station or Base in an active war zone, a country undergoing significant turmoil, or an environmentally difficult place to have a PM officer assigned to it for traditional intelligence collection operations vice a paramilitary covert action mission. Where an Operations Officer with a professional background in corporate law or finance might flounder, a Paramilitary Officer will often thrive.

Dutch was soon selected for an assignment of this type, to the kind of "garden spot" where very few traditional Operations Officers willingly choose to be assigned. Dutch was sent to man the CIA Station in ▮▮▮▮▮▮▮▮, on the western coast of Africa in 1994.

"I went back to SOG at Hqs, and was home for about six months. One day senior leadership came by, looking for volunteers to go overseas on PCS assignments. I put my hand up. I drew a slip of paper out of their hand, and it was for the Station in ▮. *Other people didn't want to go to* ▮, *anyway. So I got the assignment to go PCS to* ▮ *Station in 1994, for two years. When I arrived, the Station manning was the Chief of Station (COS), who was a mid-grade Directorate of Operations Case Officer named Bob Meehan, one female support officer and me. Shortly after I got there, the COS left the country on personal leave and told me that he was not coming back. This left me alone to handle the whole Station's roster of human agent assets. I became Acting COS. I had been handling* ▮ (a fair number of) *agents and he had been handling* ▮. (A similar number.) *Agent handling operations are demanding and time consuming, requiring considerable effort to plan and conduct them. There is no way I could securely handle* ▮ *agents.* (double my current number) *I was forced to terminate about half of them, paying them off and thanking them for their service, and telling them we would get back in touch if we needed them again in the future. I spent almost two years out there. That wasn't the greatest thing.* ▮ (the West African Country) *was and is a poor, underdeveloped place, full of crime, disease, corruption and poverty. It was dirty, dangerous and uncomfortable on its best day. It was also back in wartime again, it was just full-time war between all the different countries and factions out there. I was there during the time of the sale of blood diamonds. There was a movie made about that, called "Blood Diamond." The assignment was really hectic. Conducting traditional Human Intelligence (HUMINT) operations in a state of civil war and lawlessness is a chaotic, dangerous business. Just driving around the city is hazardous, and there were guns around every corner. That's why they liked having Paramilitary Officers take those assignments; they can cope with this type of environment better than the average traditional case officer can. So despite all the challenges, I completed my assignment and turned the Station over to a new COS at the end of my tour."*

CHAPTER 35:

THE FARM. 1996-1999 – PRESENT

Dutch was at this point sixty years old, and was a ▓▓▓▓. (A civilian grade analogous to a military O-4 Major.) His wife Cathy was beginning to experience the severe medical symptoms which would eventually end her life. Family pressures were great, as Dena and her children were also living with the family. So Dutch saw the writing on the wall and realized that he needed to be present in the home, in a safe environment which allowed proper medical care and daily attention to his family's needs. Dutch by now also had a long military and paramilitary background in formal training. He fully understood how critical it is to provide sound training based upon experience to new officers, to prepare them to meet and overcome the challenges which will present themselves as impediments to mission accomplishment. So he accepted another assignment to the CIA's storied training facility, colloquially known as "The Farm, as a Paramilitary Skills instructor.

"After that assignment, I came back and the Chief saw me and told me that he had an ongoing mission, with a PCS Assignment to ▓▓▓▓▓▓▓*, (an East African post) or I could go anywhere else I wanted to go in SOG. I had my choice. My first wife was getting sick at that time. I could either go to* ▓▓▓▓▓▓▓*, where there were no good doctors, or I could go to the CIA training base*

(known as The Farm) where there were plenty of good doctors available. So I went back to the Farm, where I had served before from 1986-88, as a Paramilitary Skills instructor. Maybe that was a mistake, maybe not. But I've been there ever since!"

CHAPTER 36:

CIA OPERATIONS TRAINING

The Central Intelligence Agency was formed in 1947, just after the end of World War II. It was chartered to collect and provide strategic intelligence for the use of US policy makers, and to carry out other activities in support of American interests and foreign policy goals. It was a direct successor to the WWII Office of Strategic Services, and many of its earliest employees were former OSS personnel. From the very start, it was understood that intelligence operations entail risks, and these operations are often carried out in inhospitable, dangerous environments. Wartime experience had clearly illustrated those risks, and demonstrated the importance of the intensive training necessary to prepare officers to execute those operations. Analysts, accountants, secretaries, computer specialists, administrative support personnel, etc, work at the CIA, plus many hundreds of other occupational specialties. Most of them never set foot outside the USA, and most of the ones who do perform their work inside hardened facilities, behind guards, walls, wire and electronic security systems, and the most dangerous thing they do on a daily basis is to travel to and from home and work. But some CIA officers incur far greater physical risks, as part and parcel of how they do their work. Operations Officers and Paramilitary Operations Officers by definition work outside the wire, in all hours, in all climates, and in all conditions of risk including unstable countries, extremely austere environments and active war zones. They often work alone, or in very small teams. And they require a high degree of training, expertise and personal confidence to be effective in their roles.

From the earliest days of its inception throughout the end of the 20th Century, the CIA's Directorate of Operations understood those realities, and tried to provide appropriate training to meet the need. Operations Officers were the predominant staffers of the ▮▮▮▮ overseas CIA Stations and Bases; they primarily were the ones doing the risky work of human intelligence collection. Not just anyone can do what Ops Officers are required to do; the work is complex, subtle and difficult. It requires a particular set of personality traits and abilities to do it well. Therefore, these particular officers were selected by a process of targeted recruiting and careful psychological profiling and screening. Once they were hired, the CIA made efforts to provide them with commensurate training. Learning about the CIA itself, becoming proficient in its bureaucracy and data systems, and becoming certified in the arcane and highly specialized art of recruiting, training, and running "human agents", or expressed less elegantly, "spies" was just the beginning of training for an Operations Officer, and many books have been written about that process. We will not rehash those specifics here.

But before an Ops Officer was deemed ready for field service, he or she had more practical skills to master. These "hard skills" were primarily taught in a now-well-known, previously "secret" training facility most often referred to by its old CIA nickname, "The Farm." The training was designed to give the new Operations Officer knowledge, skills and confidence; all three were important elements contributing to their future effectiveness. Remember that the OSS had only a few years before conducted successful and critically important operations by parachuting small teams deep within enemy occupied territories, recruiting and equipping guerrilla partisan forces, and establishing agent networks. The

1950s and 1960s showed the future pattern of geopolitics as comprising small proxy wars and covert activities instead of global war conducted by massed troops. So the Agency set about training its officers to work according to this new paradigm. And it continued to instruct Operations Officers in the skills which had been needed before, and most likely would be needed again. This meant diverse weapons training and actual bush fieldcraft and also included a Paramilitary Course, which required parachute training and cargo resupply drops. Many hundreds of Operations Officers underwent this training and put these skills to use in Vietnam, Laos, Cambodia, and other hot spots around the globe. And the primary trainers imparting these skills were military-veteran Paramilitary Operations Officers, as the subject matters were squarely within their unique specialties and skill sets.

By the late 1980's, the "hard skills" training had morphed a bit. One major change was politically driven. This type of training is exciting; put simply, it is fun! And overseas assignment was and is advantageous. It allows officers to earn extra money, live in and enjoy other cultures, and to actually perform the hardest and most fundamental work of the Agency, which in turn results in more likely promotion. The Headquarters-centric cadre of CIA officers chafed at the historical deference given to field Operations Officers, and envied them the promotions and assignments and earnings which their service (which often entails great personal risk) provided. So they began to agitate for increased representation overseas for non-Ops officers, for the ability of other career specialties to be Chiefs of Station, and to be allowed to take part in the fun, exciting training which had previously been reserved for the Ops Officers who most needed it. And the Operations Officer cadre likewise changed; under extreme political pressure to embrace the progressive dogma of diversity, equity and inclusion, CIA forcibly introduced women into the Operations Officer career path in far greater numbers than previously, when Ops Officers had primarily been military-experienced males.

The old Paramilitary Course was modified and renamed. It became the Special Operations Training Course (SOTC).

And instead of being given just to Operations Officers, it was now attended by groups of newly-hired Career Trainees (CTs) from all four of CIA's directorates (Operations, Intelligence, Administration and Science and Technology). And it increasingly incorporated female officers as trainees. Initially, SOTC training retained some of its earlier physical toughness. It required a degree of physical fitness, measured by graded performance of the Army Physical Fitness Test, with certain minimum scores required to allow participation. This was to ensure that participants could participate meaningfully and without an undue rate of injury. It featured morning exercise sessions, formation runs, ruck marches, obstacle courses, and strength-based training events. It looked a lot like military training by design, and trainees wore military BDU uniforms.

Initially (and for decades afterwards), trainees were housed in woodland squad-bay barracks, with communal bathrooms and showers. Women trainees were housed in separate barracks from the men, and organized into female-only platoons and squads, like their male counterparts. Later, after performance and attitude problems within the female-only platoons became obvious, they were integrated into squads and platoons with the males. This was the framework in place when Dutch arrived as a Paramilitary Instructor at the Farm in 1986-1988, and again in 1992.

CHAPTER 37:

THE SOTC

If you asked any US Government service officer who had served overseas ▓▓▓▓▓▓▓ (in the places where such people work) through the end of the 20th Century, they would invariably tell you that "you can always tell who the CIA people are. They are DIFFERENT." They were indeed different from their ▓▓▓▓▓▓▓ (other U.S. Government service) colleagues. Many maintain that the difference was the training that CIA provided, not the least element of which was the Special Operations Training Course, or SOTC. As of 1990, when this author attended SOTC (as all new Operations Officers did), it was still providing meaningful, exciting and confidence-building training, if the trainee chose to approach it with a seriousness of purpose. Trainees would typically begin the 6-8 week course after only a short period of about 10 weeks of Headquarters-based training to integrate them into the CIA's bureaucracy and culture. Trainees would be future Operations Officers, Collection Management (Reports) Officers, Analysts, Security Specialists, Administrative Officers, and some Technical Specialists, all of whom were being prepared as Career Trainees, and whose duties would likely in time take them to overseas postings. (In later years, timing of this course for Operations Officers would go back and forth between being attended before their actual training as Ops Officers in the Field Tradecraft Course (FTC), to attending SOTC after graduation from that arduous 6-month long certification. Instructors of the SOTC (including this author,) mostly preferred getting the students before FTC; attendees who were already graduates of the FTC often showed a lack of seriousness in pursuing SOTC training. They too often felt that as FTC-certified Case Officers with future assignments already identified, they had little to fear from approaching SOTC cavalierly.) In its transition from the old PM course to the SOTC and its introduction to a broader student population including large numbers of women officers, some of the urgency and soberly lethal intent of the curriculum was lost. It became what more than one attendee characterized as "*Outward Bound*, with guns!" Due to the large numbers of trainees who paired off and formed couples, many of whom ultimately married, it was also waggishly termed "the world's most expensive and exclusive dating service."

The premise of SOTC was simple; *You are going to serve in difficult and dangerous areas. You need to know a wide variety of skills to prepare you to do this. We will try to give you the knowledge, skills and confidence which will help you succeed and bring you home.*

Other goals were to give each student an introduction to military topics, so that future intelligence collection on these issues would be more detailed and informed. The importance of teamwork and cooperation would be made clear. They would gain an appreciation for the arduous life that soldiers and paramilitary operatives were required to endure, as they would be supporting these types of people and operations in their careers. They would learn useful skills which would enable them to do their work in hostile, austere or uncomfortable environments, and they would be inculcated with the confidence to actually go out and do the work. Finally, the training would be a powerful bonding experience, cementing the classmates as teammates and friends for the duration of their careers, and inspiring them to cooperate and support each other when needed. A simple phone call across Headquarters to an old

SOTC classmate would always be received with a desire to assist. This author believes that these goals were met in very great degree across many dozens of SOTC classes.

Attendees would arrive at The Farm on a nondescript bus, and be brought to the Training Center, known historically by its euphemistic and blandly generic title, the Field Activities Building, or FAB. This was an old brick structure nestled in dense pine woods on the far side of the sprawling training base from the more genteel buildings which house the administrative, conferencing, dormitory and schoolhouse training centers. ▌█████████████████████████████ (This sentence described the early history and construction of the building, which has long since been demolished.) It housed an impressive armory, several classrooms and staff offices. The classrooms were filled with scarred metal desks and chairs over ancient linoleum floors, with antique military training aids and the memorabilia of earlier classes displayed on the walls. The redolent scent of decades' worth of weapons bore solvent and lubricants hung in the air as a constant reminder of the long history of the base and the course. It was a place of great cultural legacy.

The students were welcomed and the basic intent and syllabus of the SOTC were reviewed. Rules for conduct and expectations for performance were laid down (but often blithely ignored by the less-serious trainees.) The students were organized in military fashion, into platoons and squads, each initially led by a military veteran student who had been selected for this task by the staff based upon a review of their biography. (Later, these roles would be rotated to allow specific students to gain leadership experience.) Decrepit 12-passenger vans were issued to each squad for transportation, with drivers designated and this responsibility rotated. The students were shown how to find the Base Dining Hall which would feed them, and the gym and recreational (bar) facility where they would spend their off-duty time if they wished. The students were then taken to a large warehouse (itself a ghostly remnant of the earliest history of the former WWII military base) and issued woodland camouflage battle dress uniforms, a combat harness, rucksack, field knife, flashlight, leather gloves, a magnetic compass, canteens and a matching aluminum cup, and other assorted field gear and clothing items. Given that the training would require a great deal of hiking, each student had previously been issued a pair of top-quality Danner boots, and told to break them in and bring them when arriving. Once their field gear had been issued, the students were driven to the barracks which would be their homes for the duration of the training. These were one-story rectangular squad-bay buildings with bunk beds and communal bathroom and shower facilities, isolated deep in the woods. Each student selected (or was left by default) a bunk, and spread out their bedding and gear. Most of the students had no idea how to assemble and fit their field gear; the military veterans in the group quickly taught the others how to put it all together, and what each piece was for. The BDU uniforms and boots were donned, and training began in earnest.

CHAPTER 38:

MRLN

The SOTC had several discrete modules, each intended to build upon the last. Map Reading and Land Navigation started things off. Many of the students had rarely if ever used a map, and few had any idea of how to use a magnetic compass. The students were taught how to read and orient a topographic map, and how to plot locations on it using a protractor and the Military Grid Reference System (MGRS.) They were instructed on the use of a magnetic compass and (after the technology developed,) a rudimentary Global Positioning System receiver, the Garmin 12XL.) They were instructed in the arts of applying magnetic variation to obtain a true bearing, how to shoot and travel a directional azimuth, and how to plot their position by triangulation on known visible landmarks. They determined their pace count, so that they could accurately estimate distance traveled by how many steps they had taken, over varied terrain. They were taken to a marked open field start point and told to traverse the field on a specific azimuth. They would arrive several hundred meters later at a line of numbered stakes; one of these was the correct destination, and they would be told if and how they had erred. Finally, they repeated this at night.

When all this training was assimilated, the students were ready at week's end for a final exercise: the All-Day Land Navigation test. Not long after sunrise and breakfast, each student was issued a military Meal, Ready To Eat (MRE) for their noon meal and each filled their canteens. They were taken to a starting point, which was a red 55-gallon barrel with its MGRS grid location stenciled on it. This was sited in the woods near one edge of the sprawling base. Each student had to plot that location on their map to see where they were. Instructors told the student the framework for the exercise: *"You are posted to a country which has just begun a civil war, and you need to get to safety. You are directed to proceed to your extraction point. You should do this discreetly, because if you are captured by rebel forces, they will execute you. You should therefore move cautiously, stay off the roads and avoid open, occupied areas. Cross roads and bridges quickly and then fade back into the woods and stay undetected."*

They were then issued a second grid location to plot. That was their first destination. They were then released individually on different specific courses to their differing destinations. One student could not simply follow another; they had to go where they were supposed to go. If all went well, they found a second red barrel at their destination, and stenciled on it was the grid location of a third barrel – which led to a fourth, fifth, etc. And the student was expected to arrive for pickup at a specific end point by a hard late afternoon time, having traversed between seven and thirteen kilometers of hills, streams, swamps, and tick-and other insect-filled wooded terrain successfully. Missing one barrel meant you could not complete the exercise. You had to find each to know where to go next. And according to the exercise protocol, if you were late, you had missed your pickup and would not be rescued.

Most students did fairly well at this. They were intelligent and had assimilated the learning points and embraced the spirit of the exercise as it was intended. But there were a few memorable exceptions long remembered by their instructors. One of these was a young woman named Lindsay: she later described her experience in a whiny tell-all expose' of her brief and undistinguished Agency service, which she publicly insists was stellar. A *Magna Cum Laude* Harvard graduate, she couldn't be bothered to take all this military skills stuff seriously, and so she quickly became lost. She resorted to walking along marked paved roads to get to her destination. (The students had been told to move to a paved road if they became lost or injured. Instructors patrolled the roads to assist and if necessary to reorient the student.) An Instructor found Lindsay walking the road, and inquired "May I help you, young lady?" She answered that she was lost, and he pointed out to her where she was, and told her to get back in the woods and back on track. Familiar with her entitled, rude and haughty personality already at this point in training, the Instructor moved a few hundred meters down the road and waited. Sure enough, out popped Lindsay to resume her walk along the road, avoiding the woods and (for her) unknown territory. The Instructor approached her again and asked "Excuse me, young lady. Why are you back on the road.?" She gave him a ration of BS about how she didn't want to go into the woods, and he quite correctly directed her to follow the exercise protocol. The Harvard Valedictorian failed to complete the exercise properly.... and formally complained that the instructor had demeaned her by calling her "young lady" repeatedly.

Another humorous incident involved a young officer named Tony. Tony was his class's screw-up. Perennially late, behind the curve and usually holding a tenuous grasp on whatever the subject matter was, he would shrug, sheepishly grin and muddle on. His classmates made allowances and tried to help keep him on track, with varying degrees of success....but this was an individual exercise. Tony quickly became lost looking for his second or third barrel. He was casting about for a while, and realized that he was in sight of the Base Dining Hall....and it was lunchtime! So he went in, was served lunch, and ate it sitting in the air conditioning, relaxing and drinking iced tea. He located the dining hall on the map easily. Then he went outside, climbed aboard a base bicycle in his BDU uniform and set off across the base roads, pedaling hard to make up time to his as-of-yet unlocated destination! Needless to say, this did not amuse the Instructor Cadre, and Tony failed this exercise.

A third instance in the Summer of 2001 was memorably funny. Not long after exercise start, I was manning a road near one of the first checkpoints. Out of the woods jogged a junior Paramilitary Officer, who was a stellar student. (He later went on to earn an Intelligence Star for heroism in Afghanistan.) Andy was a former Army Special Forces enlisted man, so map reading and land navigation was a lark for him, and he was jogging the course to try and push himself to finish as soon as possible. I verified that he needed no assistance and sent him on his way. About sixty seconds later, out of the brush popped Rochelle, one of the female students in the class. Short but gutsy, she had already impressed the Instructors as a serious student willing to push herself to meet the training's demands. She was way out of her element and comfort zone, but was trying her best and doing well. She was running like a horde of Chinese infantry was on her tail, and I thought that she might be trying to race Andy. But then she saw me and beelined for me, yelling "TICK! I'VE GOT A TICK!" We got the offending insect off of her and she continued on her way, at a much more reasonable pace. She finished the exercise in good style.

CHAPTER 39:

WEAPONS TRAINING

The following week, students were exposed to a very wide variety of weapons. This was for several reasons. Firstly, they needed to be qualified to carry the Agency's standard handguns, at the time the Browning Hi-Power 9mm or Smith & Wesson .38 Special revolvers. (As of late 2001, the Agency replaced the decrepit and obsolete Browning and Smith & Wesson handguns with Glock Model 19 9mm pistols.) They also needed to be able to understand the differences between a handgun, shotgun, submachine gun, battle rifle, assault rifle and machine gun. They would need to know what each was and be able to identify several common examples of each type, so that their future intelligence reporting would accurately describe them. What a particular unit or facility is armed with often helps identify exactly which group is being discussed, and it allows detailed mission planning by forces that may be directed against them. And *in extremis*, the officer might need to arm himself and use a locally-acquired weapon in the event of a collapse of civil order where he or she is located. This has happened several times in CIA's operational history.

The training began with handguns. The students were taught the fundamentals of safety and marksmanship, and how to load, unload, reload and clear malfunctions of their weapons. They learned how to shoot from behind solid cover, and how to draw their weapons from concealment quickly to bring them into action. They learned how to move through a building, firing at armed reactive targets when they appeared, and not firing at unarmed innocents. They did some shooting around automobiles, and in low light conditions. The officers were taught disassembly and cleaning procedures and practiced them after each day's firing. For many students, this was the first time they had ever fired a gun. Once they were qualified, they were shown and allowed to fire a collection of varied weapons including the Browning, the Smith & Wesson revolver, the Glock 9mm, the Beretta Model 92 (which was the standard US military sidearm at the time), the Sig Sauer P226, the Colt M1911 .45 ACP and the Russian Makarov PM.

The students then learned about shotguns, and were trained on the then- Agency-standard Winchester Model 1200 Defender pump 12-gauge, firing birdshot, buckshot and slugs. They also got to fire the Remington Model 870 12-gauge. These hard-recoiling weapons made both friends and enemies among the smaller students.

Next up was submachine guns. The students were taught about this intermediate class of weapons which fire a pistol cartridge, but more accurately and controllably than from a pistol. The students got to fire (and clean) the Heckler & Koch MP5, the .45 caliber US M3 "Grease Gun", the "Swedish K" 9mm (an Agency favorite during the Vietnam War), the Beretta Model 12 and the Israeli Uzi 9mm.

Training then progressed to rifles. The students were taught the fundamentals, and got to fire and clean the M16, the M14, the AK-47, the FN/FAL, the German G3 and the Steyr AUG. For the vast majority of the students, this was the first time firing most of these weapons, and most especially their first experience in firing them on full automatic. This occasionally led to some very dangerous instances of officers (usually slightly-built or terrified females) losing control of the weapon on full automatic and allowing their fire to rise above the range backstop, winging off to destinations unknown. The Instructors did our best to keep this to a minimum.

Finally, the students were introduced to belt-fed machine guns. They were given demonstrations and allowed to fire the M60, the Russian RPD, the Belgian MAG-58, the Browning M1917A6, and they were shown the firing of the Browning M2 .50-caliber and the Russian DSHK.

And to finish off the weapons module, the students were given live demonstrations and then allowed to throw an anti-personnel hand grenade and to shoot a 40mm M79 grenade launcher with practice grenades. To close out the module, they watched the demonstration deployment of a White Phosphorus incendiary grenade and the firing of a Soviet Rocket Propelled Grenade (RPG-7) usually fired by Dutch! I remember him scoring direct hits on a junked armored personnel carrier hulk used as a target on the range.

CHAPTER 40:

THE OVERSEAS PERSONAL SECURITY COURSE

The next phase of training was perhaps the most useful in everyday life. The Overseas Personal Security Course (OPSC) was all about street awareness, defensive and offensive driving. It instructed students in the art of performance driving, making the most of a car's innate power, weight distribution, traction and handling characteristics. The students spent a week learning to control a car on a driving track with various turns, obstacles, and conditions. This began with a timed run through a specified course. For many students, driving faster than they ever had before was extremely stressful. A few students were former mega-city dwellers accustomed to taking a taxi or the subway everywhere, and whose driver's licenses were still crisp and new. In addition to driving skills, students learned about ambush techniques, and how to counter them. Students were shown several live demonstrations of explosives, learning what a charge of TNT will do to a car, or a few grams of sheet explosive concealed in a letter bomb will do to a desk and a mannekin seated at it. They learned how a parked car or even a bicycle chained to a fence can be used as an explosive weapon. Actual case studies of successful and unsuccessful assassination attempts were reviewed, and students were told to be observant for indications of the pre-attack surveillance which almost always takes place for weeks or months before an actual attack. They were also taught how to inspect their vehicles before entering them each day to forestall car bomb attacks, and Instructors occasionally placed inert bomb simulators on student vehicles to drive home the point. It was a major embarrassment for a student to be shown the "bomb" under the car they had just started and failed to check beforehand, in front of their classmates. Finally, students were taught the rudimentary aspects of the Pursuit Interdiction Technique (PIT) maneuver, very close formation driving, J-turns and emergency reverses, and vehicle rams to escape vehicular roadblocks. In an intense nighttime final exercise, students were put behind the wheel of a reinforced junker car and led down deserted woodland roads into several scenarios, blinded by an opaque hood. When their instructor removed the hood, they would be confronted with a close-range attack scenario. They needed to instantly assess and react, choosing to either avoid, reverse or ram through. It was a very stressful but instructive evolution, which also drove home the point of the value of situational awareness to survival. "Keep your head up, and look around. Pay attention to what is happening. DON'T BE SURPRISED!" became an important habit for the officers who took the lessons of this training on board. On the last day of the course, the students ran the same timed driving course as on the first day. Usually, their time had improved by several seconds, and invariably, their degree of control was considerably better than before training. Agency records document that numerous lives have been saved across many decades by this outstanding training course, both overseas and here in the USA.

CHAPTER 41:

MARITIME OPERATIONS

This module began at the Base swimming pool, to determine who in the class actually could swim, and to provide some basic water survival instruction. Classroom instruction on Maritime Operations included tides and currents, marine navigation, small boat handling, basic outboard motor starting, operation and maintenance. Instructors taught the students how to launch and recover a trailerable boat. Daytime and nighttime river and shallow creek runs introduced students to the practical aspects of driving a small boat from Point A to Point B successfully, knowing where you are and getting to where you want to be without running aground or into anything. This skill is useful when the place you are is no longer viable, roadblocks are in place and a body of water is nearby, with boats available for hire or theft. A little bit of knowledge can make an officer capable of "getting the heck out of Dodge" safely with his family, colleagues or agents in tow when their environment has turned hostile and "anywhere but here" is the order of the day. A final exercise is fondly remembered by most SOTC students. Teams were issued radios and embarked on small boats with outboard motors (sometimes in inclement weather.) Each team navigated all day ▆▆▆▆▆▆▆▆ (in a riverine environment) to a distant marina for recovery at the end of the approximately 50-mile journey. This was great fun; a day boating on the river with your friends (unless it was raining or cold, in which case the fun meter indicated somewhat below maximum.) This was also not a trivial undertaking, and successful completion resulted in tremendously increased confidence.

CHAPTER 42:

GROUND OPERATIONS

With all those fundamentals out of the way, demands on student performance accelerated. The students were taught the basics of small unit tactics, formations and movement on patrol. They learned and practiced emergency medical first aid. Students learned to rappel off a 34-foot high rappelling tower, with a sloped face, a vertical face and a simulated helicopter skid. This introduction to acrophobia was a major life learning and confidence building event for some. They learned how to move through the woods quietly, carrying weapons in proper patrol order, and crossing roads, water hazards and bridges in a tactically sound manner. There was a two-day wilderness survival module, taught in a remote area of the base, where improvised shelters, fire building, water harvesting, simple snares and live foods preparation were taught. After being given a demonstration of killing and preparing a snake, a rabbit and a chicken, small student groups were then given a live rabbit and some wild turnips for their dinner. (Some ate, many didn't.) Training exercises during this module might, for example, consist of giving a student squad the mission of patrolling to a specific location in the woods, where an "enemy" target facility was located. The student unit would have to use their map reading and land navigation skills to plot and locate the target. They would approach it as an armed patrol (with weapons modified to fire blanks only) and get close enough to collect useful intelligence about what was there. The "facility" might be a clearing in the woods, with old military hardware like a Cobra helicopter fuselage, a junked M4 Sherman tank, numerous fuel drums, an anti-aircraft gun or fake bombs or missiles present, guarded by instructors armed with blank firing weapons. Students had to get close enough to sketch or take notes on what was there, in as much detail as possible without being detected. A final exercise involved dispatching small groups of two or three students by inflatable boat up a creek to a remote area of the base. This was again couched as an escape and evasion exercise, in a country where they would be killed if caught due to civil instability. They would have to wade ashore in thick riverine mud, evade into the bushes and discreetly proceed several kilometers to a specified distant map coordinate/extraction point the following day. When darkness fell, the students needed to establish a patrol base and bed down in the woods for the night. Instructors roamed the nearby roads and woods, putting on a show of firing illumination flares and blank ammunition and sounding bullhorns and sirens, simulating a search for them. For many students, this was the first time in their lives they had ever slept (or NOT slept) outdoors.

This module put several of the student's new skills to use in immediately practical fashion. It was also physically strenuous, and it was the point in training where individual shortcomings became readily apparent. It was why physical fitness requirements as measured by the Army Physical Fitness Test were assessed at the beginning of training, excluding some weaker or medically unfit students. Patrolling silently while carrying a rucksack, weapon, spare magazines, water, and other needed equipment across kilometers of hilly woodland terrain is hard work. It is hot, sweaty, uncomfortable, tiring and unforgiving of personal weakness. For the military veteran students, it was something they had probably experienced before, to some degree at least. But many students had never done anything like it. They did not see it as relevant, they were not physically prepared to do it, and they hated being made to do it poorly in front of their peers.

Female students in particular demonstrated issues in this area. When women first began to attend SOTC in greater numbers, they were initially segregated into female-only platoons. As most or all of the women had little relevant similar experience, their platoons invariably under-performed behind their military-experienced and stronger male counterparts. Female platoons were always issued M16s, a relatively light weapon, and they had great difficulty carrying an M60 machine gun and associated ammunition belts. Their male counterparts carried M14s or FALs, much longer and heavier rifles and managed the M60 effectively. The women simply had extreme difficulty carrying the weight, going the distance, making the time limit, and moving quietly while uncomfortable. And grasping the incongruity of their trying to perform these difficult tasks, they often displayed a lack of seriousness about the training. A couple of legendary examples illustrate this. In one case, a pizza delivery vehicle approached the guarded front gate of the training base, and told the guards that someone had ordered several pizzas. Out of the nearby woods scuttled a camouflage-clad female trainee, who paid cash and disappeared back into the brush bearing the evening meal for her patrol. The Instructor cadre were not amused. A second incident was even more awful in its implications. While the students were on an overnight "don't get caught or you will be killed" patrol exercise, the temperature dropped and cold rain began to fall. Rather than spend an uncomfortable night out in the weather (which their male counterpart platoon did without a second thought,) the female platoon marched to the Base's front gate and "surrendered" themselves to the gate guards. They preferred captivity to discomfort. Shortly after that, female-only platoons were abolished and women were integrated into male platoons. This had three benefits: It was oh-so-politically correct. The men could help carry the females along towards mission accomplishment. And male students would brook no stupidity like surrendering en masse when things got tough.

CHAPTER 43:

AIRBORNE OPERATIONS

Finally, as a culmination of all that had gone before, a final APFT was held, with higher-than-minimum physical standards required. And those students who passed were allowed to participate in Airborne parachute training. This was actual, individual static line parachuting, not a tandem parachute jump while attached to an experienced skydiver. The latter is a recreational lark, akin to riding a roller coaster; no skill or knowledge is required. A bag of potatoes could do it. The former requires an active, willing jumper. Why, you may ask, did CIA feel it was useful to train officers to jump out of airplanes? For several reasons. Remember, only one generation ago in WWII, the OSS had to parachute officers into occupied Europe to conduct operations. Agency history is replete with other parachute operations carried out by officers and agents. It is a useful skill when the operational demands are critical and business-class airline travel is not an option. Secondly, Airborne training emphasizes attention to detail and proficiency. It causes the student to focus COMPLETELY on the task at hand, to listen carefully to instruction and to master the material. It is a useful disciplinary tool in this way. Finally, it causes a trainee to face, manage and overcome a very real, very rational degree of fear. It forces them way beyond the comfort zone of any normal person.

It makes them function despite their fear, carry on and accomplish a task. And once achieved, it provides an unparalleled degree of growth in personal confidence. When faced with a scary situation in the future, the Airborne qualified officer is likely to look deep inside and say to himself "this isn't so bad. I jumped out of airplanes, I can do this." And that confidence is priceless.

The Airborne module included airborne materiel delivery, with students preparing and rigging supply equipment bundles for air drop (under the tutelage and supervision of qualified riggers), marking a nighttime drop zone, and ultimately kicking the bundles out of an open door aircraft in the darkness, with other students receiving them on the DZ like a partisan group being resupplied with rations, ammunition and other needed materials. Then jump training began in earnest. The same techniques as taught at the US Army Airborne training school were drilled into the students, on similar apparatuses as are used there. An old C-130 fuselage hosted instruction on how to stand up, hook up static lines, perform last-minute equipment and buddy checks, and to approach and depart the aircraft door. A sawdust pit with platforms of varying heights was the scene of the torture of learning and repeating countless Parachute Landing Falls (PLFs) in every direction. A Suspended Harness Trainer taught how to steer the canopy. A Lateral Drift Assembly taught additional canopy control techniques. And numerous trips on the 34-foot jump tower required students to correctly don a parachute harness, hook up to a pulley on a wire, jump out, vertically drop ten feet or so, and slide down the horizontal wire to a tall berm at the end, under the watchful eye of the Instructors who called corrections to form and simulated malfunctions requiring corrective action. When all was ready, the students boarded a DC-3 aircraft and made a total of five static line parachute jumps including one with combat equipment, the same as required at the Army Airborne School at Fort Benning, GA. And those who completed this training were presented with a certificate and a pair of silver basic parachutist's wings by the CIA's Deputy Director for Operations. Not all students earned those silver wings: injuries took some out of the class, and some timid officers quit before jumping or after one jump. But those classmates who completed the training together shared a career-long bond and a life-changing experience, exactly as intended. One CIA Operations Officer who completed Airborne training in 2001 under Dutch's instruction described his experience this way, writing a letter to his then-teenaged daughters:

"13 August 2001

AIRBORNE!

Today I made my first three parachute jumps. It is such an unusual experience that I wanted to share it with you, to let you see, even if only vicariously, what it is like. I also wanted to get my own feelings down now, while they are still fresh, so I can remember always what today meant to me.

My class has been training for some time now, preparing for this big step. We have put in long hours over many days, in grueling heat and humidity, learning every facet of what a parachute is, how it works, and what we are expected to do to control it and to make it function properly. We have performed dozens of practice parachute landing falls (PLF's,) jumping from various platform heights and throwing our entire bodies onto the ground so as to distribute the shock of landing over our entire musculo-skeletal structure rather than taking it all on our relatively fragile legs. We have tirelessly drilled full and partial malfunction procedures, canopy control, water landings, tree landings, power line landings, 'chute collisions, towed jumper procedures, and actions on the aircraft and in the door. We have made countless door exits and ridden a 34-foot tower trolley repeatedly to demonstrate all of these skills. Our instructors, both military and USPA free-fall qualified, have literally 10,000 jumps between them. Many are combat veterans. They feel we are not only ready, we are trained to a higher standard than the Army Airborne course at Ft Benning, GA requires. We have trained more technique repetitions on the same apparatuses as are used at Ft. Benning. Four class members verify this fact to the rest of us, as they are former military and prior Airborne graduates. The pain of our many aches, scrapes and bruises fades. We are ready. We know it. It is time to go jump.

*The day is bright and sunny, but storm clouds are approaching and the winds will increase later. We suit up and get through two JumpMaster Preparatory Inspections, ensuring that our rigs are donned correctly and in ready-to-use condition. The butterflies have started for some class members (okay, **all** class members.) A few worried looks are being exchanged, and the time seems to be dragging. Some classmates are disappearing into the bathrooms. I just want to get on the plane and do it. I am pleased to be jumping with a great group of student colleagues who I respect and like a lot. For many of them, this is a lark, a chance to do something fun which they will never do again. For others, it is a rite of passage, something they are expected to do and so they are gritting their teeth and getting it over with. For me, it is the culmination of a goal I set over twenty years ago, and that I have never let go of. I want to complete this training, and must do so without getting injured en route and having to start all over again. My ribs hurt quite a bit from the impact of many landings during training. I am almost 41 and no longer immortal, nor made of elastic, quick-to-heal rubber as I was in my youth. I want those silver jump wings badly.*

As an instructor here, I have asked for and been given the honor of being the first man in the first stick – the first of our class to exit the door. We board the aircraft in inverse order, with me last aboard. A beautifully well-maintained and -preserved DC-3, this aircraft is blessed with a wide, tall exit door and a wonderfully slow forward airspeed. We sit facing inboard along the fuselage, on plastic seats, and attach regular seat belts. Leaning into the forward thrust, we taxi, and lift off. The door is open, and I am given a bird's eye view of the terrain and wide river whipping past below as the noise level and breeze in the plane rise.

All too soon, the plane reaches 1250 feet, and the Jumpmaster gives us the sequence of commands putting us into motion. We stand up, hook static lines to the steel cable running along the inside of the aircraft, and perform one last equipment check. Mine is checked by the person behind me; total trust in each other is required, as I cannot see my back and the vital static line which will deploy my parachute. I know my friend and colleague is doing it right, as we have trained together.

I am given the command to "stand in the door!" I pivot right, placing my right foot toes hanging over the door sill, with my left foot back and coiled to thrust as I am semi-crouched down. My torso and head are erect, my arms straight and fingers extended onto the outside of the door frame on each side, against the cool skin of the plane as the slipstream whips by at 100 mph. I look, not down at the ground, but at the horizon as I tick off the eternal seconds waiting for the jump command and mentally review what my actions must be to give me a clean exit and opening. I am coiled to spring, with all my colleagues' eyes on me, and I know well that there is to be no hesitation, there is no going back. Life is about to change forever.

The Jumpmaster slaps my lower leg and yells "GO!", and I leap out into the nothingness of air over a large creek and a forest of trees. I snap into a good body position and begin to count "one thousand, two thousand..." My feet and knees are together, legs straight, body slightly bent at the waist. My hands seek and cover the ends of my reserve parachute, fingers splayed, my elbows locking tightly to my sides, and my chin drops onto my chest. The slipstream has instantly snapped me into a 90-degree sideways position and also rotated me to face the tail of the aircraft. It has suddenly become much quieter, and I feel the shock of deceleration as my canopy opens above me. I finish my count of "four thousand," *reach above me and take control of my canopy steering toggles, visually inspecting my 'chute for any malfunctions. Thankfully, there are none, as a beautiful symmetric circle of OD nylon has formed above me. I locate the drop zone, and turn into the wind, which is indicated by a billowing yellow smoke grenade far below me. A white pea gravel pit at the DZ's center makes a white dot down there, and the vehicles parked around it look small. It is so quiet, I can hear the joyous whoop of one classmate who is both much smaller and lighter than me, and who exited after me, but who has somehow managed to get below me in flight. I can hear the conversations of the people on the drop zone.*

I make a few canopy turns, steering onto the clear, green field that is our drop zone, the ground comes ever closer, and I concentrate on keeping the canopy pointed into the wind so as to minimize my horizontal speed for landing. As I drop below the tree line, the ground comes up much faster, and I assume a good landing position. I make a pretty darn good PLF, and hardly feel the shock of landing at all. I grin like a fool as I run around the apex of my still-billowing canopy to collapse it and prevent it dragging me along the ground. I am elated beyond words. I have made my first jump. I stow my 'chute in a kit bag and walk back to the DZ center.

I make two more jumps before the day is through, and my technique and landing accuracy improves with each one. We have had to wait for a couple of hours as the rain has come in sheets. Finally it clears and we go back up for our last run of the day. On this last jump, a close friend and instructor colleague jumps right after me, and we fly down together in a light rain and with overcast clouds. His first jumps were nearly 30 years ago, while he prepared to go to Vietnam as a young Special Forces officer. He returned from there with two Bronze Stars and a Purple Heart, awarded in recognition of a serious gunshot wound to the leg. What an experience I now can share with him.

*Whatever else I may do in this life, whatever adversity I may face, I am now better prepared for it because I know that I have faced hazard and perfectly rational fear and functioned well despite both of them. It has been said that true courage is not the absence of fear, but rather taking action despite fear because you judge that something else is more important than that fear. After my first jump, the Drop Zone Safety Officer, a grizzled old former US Army Master Sergeant who I have known for 11 years since he was one of **my** original instructors, comes over and gives me a hand shake coupled with a bear hug and a "good jump, guy-congratulations." I am a bit old for hero worship, but **this man is special**; interned in a WW-II Japanese prison camp in the Far East as a child, he later joined the Army and was a respected member of MACV/SOG in Vietnam. He holds the Silver Star and Purple Heart medals. (In case you don't know of SOG, they performed **extremely** hazardous missions behind lines in Laos, Cambodia and Vietnam, and took the highest casualties of any US Army unit ever.) He is a soldier's soldier, and a living icon in our business. If this man feels proud of me, I feel justified in being a little bit proud of myself.*

All of my instructor colleagues make a point of coming over to congratulate me. I am blessed and honored to be accepted as a peer into this company of truly courageous and dedicated men. After two more jumps tomorrow (one with a 40-lb rucksack attached to me,) I will wear the silver wings of an Airborne parachutist with pride. Whatever else I may be in this life, from then on, I am Airborne!"

That Instructor was me.

Unfortunately, the story of SOTC does not end happily. In about 2002, the Agency decided that SOTC was an anachronism. Several officers who had hated SOTC when they were forced to attend it and who had chafed under the memories of their personal shortcomings as students in it had risen to be Chiefs of Personnel for several Agency components. They took exception to physical fitness standards excluding fat, weak, or medically unfit students from training. They considered this practice discriminatory, and wanted to protect younger colleagues like themselves from the demands of participation. They questioned the relevance of the subject matter (as if fundamental life skills and personal confidence can ever be irrelevant for an Intelligence professional.) So they convinced the Deputy Director for Operations at the time (an old genteel-Europe hand who is best remembered for his undistinguished operational record plus some highly inappropriate personal conduct in an underground parking garage at Headquarters) to kill the program. And CIA no longer prepares its Operations Officers for careers in difficult or hostile environments in this way. CIA Operations Officers used to be trained in these skills, routinely. They are no longer. The CIA and America are weaker as a result.

If you still question the relevance of SOTC as it was, let me tell you about my friend and student, Helge Boes. Helge was a US citizen, born in Hamburg, Germany. He was brilliant, well-educated, multi-lingual and extraordinarily talented as a CIA Operations Officer. He had a degree in Political Science from Georgia State University, and a Juris Doctor degree from Harvard Law School. Helge joined the CIA in January of 2001. He was a stellar student in all of his operations training, including the SOTC, where I met him as one of his instructors. After the terrorist attacks on America on September 11th 2001, Helge wanted to get into the fight against Al Qaida. He volunteered for service in Afghanistan in the spring of 2002. Helge had no military background, and little hard skills experience beyond what the Agency gave him through SOTC. But given his abilities and the needs of the time, he was assigned to a ▇▇▇ unit of ▇▇▇▇▇ CIA Paramilitary officers in Afghanistan, to contribute his expertise in agent recruitment and handling operations. By all accounts, he was a highly-respected and well-liked member of the team, which was operating in spartan battlefield conditions, living out of rucksacks and carrying weapons at all times. This is precisely the environment that SOTC was originally intended to prepare an officer for. On February 05th, 2003, Helge was in an austere and remote area of Afghanistan with his team, participating in a live-fire training session, where local security forces were being taught to use hand grenades. He was killed when a foreign-made grenade exploded prematurely very close to him. Unfortunately, Helge had never had the opportunity before that day to see a properly-run grenade training range or session. His SOTC Class did not experience that valuable training exposure. Some Headquarters bureaucrat had decided that throwing grenades in training was far too dangerous an activity for modern CIA officers to engage in, and that such knowledge was irrelevant to their duties. So it was cut out of an already vastly watered-down SOTC curriculum. Had Helge ever seen and taken part in a properly-run grenade pit training session, might he have been standing somewhere else that day, or might he have helped design the training so as to be conducted in a safer way, avoiding the casualties that resulted? Might he still be with us? It is hard to say. Helge isn't here to tell us. But in my opinion, the CIA gave him less training than he deserved before his deployment to the war zone. Someone behind a desk in an air-conditioned office in Headquarters decided that he didn't need it. I disagreed then, and I disagree now. And America lost one of her brightest sons.

CHAPTER 44:

RETIREMENT. SORT OF.

In 1999, Dutch had been with the Agency for thirteen years, and he was sixty-three years old. He was at a plausible retirement age, and he decided that additional field deployments were not likely for him. He had reached the grade of ▉▉▉▉ (a civilian grade analogous to a military O-4 Major) in his time at CIA, and was still serving as an Instructor at the Farm, training new generations of CIA personnel for overseas assignments. Dutch put in his papers to retire from active CIA staff employment. Of course, his SOG ▉▉▉▉▉▉▉ colleagues and CIA management all wanted to send him off with appropriate recognition for his dedicated service. A formal retirement ceremony at Headquarters was held for him, attended by many of his Agency friends and colleagues. Others sent their regards from their far-flung overseas postings. As at most of these types of celebrations, speeches were made, refreshments served and farewell gifts presented. Some of Dutch's exploits were mentioned, and one particularly humorous speech recalled that when Dutch left the Army in 1986, he had initially been uncertain about what he was going to do next. He had about decided that a second career in the U.S. Postal Service might be a comfortable, steady job! The notion of one of America's most experienced and elite covert warriors delivering mail in short pants for a living elicited a burst of laughter from the attendees.

CIA's traditional retirement gift is an American flag and a personalized clock – symbolic of the time spent in a career, standing watch in America's defense. That clock is proudly displayed in Dutch's home. Dutch's SOG ███████ colleagues also presented him a beautiful Colt 1911A1 .45 pistol, in a custom frame for display. It lives on the wall of his study, reminding him daily of his career, his many adventures and his beloved friends and colleagues. As this was and is a fully-functional pistol of the same design as the trusted sidearm he carried into the jungles of Vietnam, Laos and Cambodia, it would have run afoul of the CIA bureaucracy's regulations against bringing weapons into Headquarters. His SOG friends had therefore deactivated it by removing its firing pin. It was (and is) not capable of firing as configured. Knowing Dutch, however, and "just in case of future need" – they also passed the firing pin to him. He still has that critical part, too. "Just in case."

But the Agency was not yet ready to let Dutch go. CIA offered him a contract to continue to work as an Instructor at the Farm as an Independent Contractor. Never one to fail to volunteer if he is needed, Dutch of course accepted, and continued to work in that capacity – for twenty-three more years.

CHAPTER 45:

PASSING – AND A NEW JOY. 2006 – PRESENT

As mentioned in Chapter 18, Dutch's first wife Cathy suffered for many years with a series of debilitating illnesses. She was an insulin-dependent diabetic, and later, exhibited severe symptoms of Chronic Obstructive Pulmonary Disease (COPD). Dutch and Cathy sought treatment for her from a series of doctors, and he provided her with the best support and care he could, while remaining at home and working as an Instructor at the Farm. Unfortunately, these serious illnesses ultimately ended Cathy's life, and she passed away in 2006. Dutch took her home to North Carolina, to be buried near her friends and family members.

Dutch was for the first time in forty-two years, a single man. As was his way, he grieved privately and carried on with his work supporting the training of CIA officers. But providence was about to open a new door for him, in the form of a loving, kind, and gentle woman familiar with the long struggle he and Cathy had endured.

Kathy Dodson Wierenga tells the story of their meeting: *"I helped open up a new concierge medical practice. I was the concierge, or membership director for the practice, so I had* duties like I had always been doing in the hotel industry. There were only two doctors, and we wanted to limit the practice to two hundred patients. The patients had access to the doctors 24 hours a day, and they paid a fee to do that. Dutch called one day, and was interested in putting in an application for membership for his wife Cathy. I took an application over to his house, and she joined. He would bring Cathy over to the practice for treatment by the doctors. She was a patient of ours when she passed away. Our doctor was over at Dutch's house at two in the morning when she passed. I was over there by seven in the morning, bringing food and necessities. In my mind this was a part of my job description, I did everything. We had an Office Manager who was also an Emergency Medical Technician. She kept a radio on the desk, and we could hear when our patients were going into the hospital. The Emergency Room staff all thought I was a doctor (laughing). I was always super dressed up, and visiting the ER frequently. I could go to the ER and ask "what room is so-and-so in" and get access. I wasn't going to see the patient, I was there for family members. I'd ask "do you need a coffee, or a Coke? Did you leave the iron on at home when you left, do you need anything taken care of? Do you need me to feed the dog?" I had a guest call me one day and he was lost, he didn't know where he was. I was able to help him get oriented. I just did stuff like that.

I left a package of medicine outside Dutch's door one day – he thought it was a bomb (chuckling.)

(Dutch interjects with a laugh: *"I almost threw it out!"*)

So I knew him a little bit. I kept trying to call him "Mr. Wierenga" but couldn't pronounce it right. Finally Cathy told me "just call him Dutch, everybody does!" She never did – she always called him Jan. She passed away in August 2006, and he took over her membership. So he was coming to the practice for routine medical care when he needed it. When you went in, you had an hour for an appointment, even if you were only there to get a shot.

So the guys especially would sit and talk to the Doctor forever. He did that. Well, we got bought out by Riverside Medical and they decided to close us down. This was in April or May of the next year, 2007. So he asked me and the Office Manager to go out to dinner with him. He wanted to thank us for taking care of Cathy. At the last minute, the Office Manager couldn't go. So I met him, and we went to this little Mexican restaurant near his house. I knew Cathy, I had met Dena. We sat there for four hours, talking about everything. (Dutch wryly remembers: "She squeezed my hand until it turned blue.")

He was like "this is really nice, maybe we can do this again sometime!" I said OK! So we just started seeing each other. I didn't even tell him where I lived for two months. (Dutch chuckles: "yeah, I kind of wondered where she lived!") *I figured he had seen my license plate number and I knew who he worked for; I thought he could figure it out. So we started just meeting for dinner, different places. After a couple of months, we kind of started dating. In October we started talking about getting married. I'd never been married before. On the 03rd of November, we went to this little cabin we had been going to, up in the mountains in Love Gap, Virginia. The doctor who introduced us was his Best Man. My family had all passed away, my sisters, and my Mom and Dad, but my cousin was my Maid of Honor. We rented a bunch of cabins at this little place, they didn't have TV or anything. We loved it. We got married on the porch of a chalet, with a beautiful mountain view behind us. That morning it had been freezing cold, and we didn't get married until about two o'clock. The sun came out and the leaves were still on all the trees, it was absolutely gorgeous. Our decorations were hay bales and pumpkins, so it was nothing fancy but it was great, we had a great time.*

People who have worked with Dutch always tell me "He's a legend!" And I'm like, okay, tell me why! But of course, they can't."

Dutch and Kathy are obviously very much in love. They literally beam at each other when in the same room. Dutch explains that they plan to be buried together at Arlington National Cemetery when the time comes.

CHAPTER 46:

THE VERSTEEGH FAMILY, AND DUTCH TODAY

Given all that they collectively and individually went through, it is fairly miraculous that all the members of Dutch's immediate family survived World War II and went on to live fairly normal, mostly happy lives. Not surprisingly, however, the stress of those turbulent years took its toll on all of them.

EMIL AND CAROLINA: Dutch's father Emil and mother Carolina left Indonesia together on a passenger ship in December of 1954. They arrived back in Holland in early 1955. Emil and Carolina found that the cumulative effects of their traumatic experiences, and the new and modest realities of their reduced economic circumstances had driven them far apart. The marriage did not survive, and they obtained a Catholic annulment and divorced in about 1956. Emil ultimately remarried to a third wife who was a nurse and he worked as an interpreter for the Coca Cola Corporation. Dutch had the opportunity to see his father again in 1977; then a fairly frail man of 76, Emil was still driving himself (albeit quite poorly) in his own car. Dutch's sister Millie arranged for Emil to come to her home while Dutch and their mother Carolina were visiting with her during a period of leave from the

US Army. Dutch was able to make his peace and to interact with his father one last time before his death. Emil died in 1985 at age 84 of prostate cancer. He refused to seek medical attention and collapsed at home. He was cremated and his ashes were scattered on the ocean near The Hague.

Likewise, Dutch's mother Carolina lived out a long life, passing away at age 81 in 1987. She is buried near The Hague, in a cemetery plot given to her by a friend.

JOOP: Dutch's oldest brother Joop emigrated to Canada after the war, and had a good life there. He became a very successful Canadian Air Force nurse, who Dutch insists had natural powers of healing. He ministered to a wealthy man in Canada, who left him a nice house, car, and financial bequest when he later died. Joop married and had a family, and was living a very comfortable life in Canada in retirement. But his wife wanted to move back to Holland, so he emigrated back there in time. He died in 1996 in Holland.

MIEL: Dutch's elder brother Miel lived a very interesting life of his own. Captured and taken to Japan in a cargo ship in 1942, Miel miraculously survived unspeakably harsh treatment and conditions. He was held in a prisoner labor camp near Nagasaki, Japan. He witnessed the explosion of the second nuclear weapon which was dropped on Nagasaki on August the 9th, 1945. The *Fat Man* device detonated at 11:02 local time at an altitude of 1650 feet, with a force of 21 kilotons of TNT. It instantly burned over 14,000 homes, and eventually killed well over 100,000 people due to its combined effects. Miel saw the historic explosion from a safe distance, suffering no ill effects from it. He spent a total of five years in Japan (1942-1947). Miel emigrated to Australia in 1947 at age 20, and became a member of the Australian Armed Forces. Circa 1948-49, Miel accompanied Australian forces into Papua New Guinea as part of a program of territorial exploration, engaging and documenting primitive New Guinean tribes they encountered. During this expedition, he was forced to eat part of the hand of a human being; the tribe which was hosting the group at the time was still practicing ritual cannibalism, and served up the body of a recently-killed enemy who they had dispatched in inter-village battle. According to their tribal custom, this was a high honor, and to refuse it likely would have provoked the massacre of the Australian delegation. The members therefore did what they had to do. Miel was by this time a very well-developed young man, capable of amazing feats of strength. Miel habitually carried a large staff as a weapon. He married a Dutch woman and one of his first children was born during his time in New Guinea, but he ultimately returned to live in Australia, where he had a total of six children including a medical doctor. One of the children is an anthropologist, and visited Dutch's sister Millie in 2020. After leaving the Australian armed forces, Miel took a job as custodian (janitor) of a public school; due to his hard work and innate intelligence, in time he rose to be Superintendent of the school! In about 1959 while he was working in the Dutch government as a minor functionary, Dutch met another Dutch government official who had actually met and befriended Miel while assigned in Papua New Guinea in the late 1940s. This man showed Dutch several photos of Miel as a young man in New Guinea, including one memorable snapshot with Miel lifting the front end of a Jeep. The man confirmed Miel's incredible strength and personal abilities, providing Dutch some very much appreciated context to his brother's post-war life history. Dutch never got to see Miel in person after his capture in 1942; he had put in a leave request to go visit him in Australia in September 1969, but was wounded on the SOG mission which almost took his leg and his life. He never got to go to Australia. Miel died of Parkinson's disease in 1992 at age 65. He is buried in Australia.

ONNO: Onno performed military service as a private in the Dutch Army after World War II, remaining in Indonesia for some time. He was a motorcycle courier in the Army after the war. He was happy roaring about on a big Harley Davidson motorcycle. Typical of his personality, Onno drifted about quite a bit. He left the Army and became a coal miner in Europe. He later became a waiter, working in an Indonesian restaurant in Holland where the waitstaff wore native dress. Dutch's sister Millie and her husband unexpectedly encountered him in this fashion, wearing a headdress and sarong as he waited tables. He finally attended formal schooling and graduated with a certificate as a jet engine mechanic. He worked on fighter jet engines for the Dutch air force. Unfortunately, he certified an engine he had just worked on as ready for flight, and the plane crashed on subsequent takeoff. The guilt greatly affected him and he descended into dysfunction and poverty, ultimately relying on Dutch government public welfare assistance for his sustenance. Both Onno and his wife suffered from ill health and both were on oxygen during a visit by Dutch late in life, even though both still heavily smoked, even when actively breathing concentrated oxygen. He died of COPD in The Hague in the late 2000's.

MILLIE: Millie is as of 2021 ninety-one years of age and is still living in Holland. She was married to a KLM pilot after the war and accompanied him there. A vital and spry nonagenarian, she still goes to the gym and exercises. Millie was always emotionally very strong, and managed her household and family finances as a very bright, capable woman. She helped provide some of the details and date corroborations for this book.

PATRICIA: Patricia is as of 2021 ninety years of age. A widow, she now lives in New Mexico, but is unfortunately in failing health. Artistically talented and sensitive, she has lived the lifestyle of a bohemian community artist and has had several successful showings in galleries, but didn't want to sell any of her work. She still has long straight hair, belying her Indonesian ancestry. She remains emotionally devoted to Dutch, still calling him her "little broeder." I imagine that for as long as her deteriorating memory endures, she will always think of him as that little boy on the trail, holding her hand as they are fleeing the Japanese soldiers.

DUTCH: Dutch today is a spry 85-year old man. Until Summer 2022, he remained fully engaged in support of training the latest generation of CIA officers, giving them his best if they would but pay attention. He is now fully retired. His eyes are bright, and his memory of many details is still quite impressive. He has been troubled by symptoms of the prostate issues which ultimately killed his father, but he has sought treatment and there is no reason to believe that he will be leaving us any time soon. Dutch enjoys a very happy marriage with his wife Kathy, who obviously adores him and takes wonderful care of him. They are looking forward to some post-retirement travel about the USA and perhaps the wider world. Dutch has always been a wonderful colleague and teammate. He would literally give a friend in need his last dime or the shirt off his back if he felt they required it. Dutch remains quiet and low-key in personal interactions. He is kind, thoughtful, and still possessed of a strong sense of humor. He has a smile and a polite greeting for everyone he encounters, whether they deserve them or not. Dutch is especially proud of his long service to America in many difficult circumstances, across six decades. He invariably wears a Special Forces ball cap, further adorned with a MACV/SOG unit crest, his Master HALO jump wings, and small Silver Star, Bronze Star, Air Medal and Purple Heart ribbons. These days, very few of his fellow Americans have any idea what those prestigious symbols mean, or of the costs with which each was earned. I have heard it said that one of the greatest indignities of advanced age is that younger men cease to think of you as being dangerous. Dutch Wierenga is to his core a professional soldier. He will never suffer that fate.

CHAPTER 47:

DUTCH'S SUMMARY

This year (2022) I'll have 60 years in service. I guess I'll stop then. My wife Kathy and I have only been married twelve years. I thought we would have some good time and enjoy life a little bit more, while we still can.

SPECIAL FORCES TAKEAWAYS: *I learned to work with different people, a lot better than the regular Army. Things are done a little bit differently, and better, than in the Regular Army. Everything is done in a lot better way. There are some sputters here and there, but in general they're a lot better. There are a lot of wannabes these days, but from the wannabes to the have-beens, I've got a lot of respect for those guys. It's a lot different. I'd do it all over again. If they would take me! No doubt in my mind, I would do it all over again. It's just one of those things. Other people have to take over. But they're doing pretty well! And I'm glad that the President is calling people back from Afghanistan. We've already lost about 12 CIA people there. We should have gotten out of there when we got everything tamped down. Those people are going to live the way they want to. We aren't going to build a society where they live like Americans would do. Now that the Americans have left, they will fight each other again. They're fighting each other now. But the war's been going on* *for what, twenty years now? It's a hell of a long time. Luckily, I'm hearing that we (the USA) haven't lost that many people compared to ten years in Vietnam, 58,000. Only about 2000, or something like that. In our generation's terms, it's not that bad, compared to our times. I don't know. There are some other places that we could probably help out more, but they're not asking my opinion. If we put a thousand people somewhere for a few months, and get some of them killed, it doesn't do anything. But that's the President's decision. But the withdrawal from Kabul was horribly mismanaged. I was alive to see the fall of Saigon, and the fall of Kabul, and Kabul was much, much worse. Really there was no comparison. It was just totally incompetently done. I was so mad, when it was happening, I wanted to go over there and help get our people out. President Biden and his people just totally blew it. They were trying to work President Trump's objective, but they did it their way, which was totally wrong. President Trump was going to keep the troops there in force until we got all our people out, then pull out the troops. Biden tried to do it the other way around! And the abandonment of Bagram Airbase was the stupidest thing.*

CIA SERVICE: *I've taught a bunch of subjects, including weapons, shooting, counter-attack defensive driving, surveillance detection. I enjoy teaching. Especially when people come back and tell me that my training saved their lives. One of the guys that was killed in Benghazi...* ▇▇▇▇▇▇▇▇▇▇ *(naming the deceased officer)... he was one of my students. But you can't say anything about it. Just the reaction of people, bad reaction. We sent them over, they didn't have any experience, we sent them over anyway. It's just the way things are. He just didn't have any experience, didn't know what to do. It's bad, but there's not much you can say. It happened. That's the way things happen in the real world, every day. Right now, I've got quite a few former students who are doing very well. As a matter of fact, one of them was very high in the Agency, serving directly under former Director John Brennan. I don't think I would want to do that.*

I really appreciate that every week when a new course started, former students asked if I was still there. They wanted to see me, but in my role, I had to be invisible until the training ended. It's a good thing, to see people come back, and they're doing well. I hope I have had a positive influence on a lot of people. The worst one I've ever had was an old guy who came back and said he was eighteen when I was his instructor. That was decades ago! He's still alive, too.

There aren't many instructors who can teach what we teach today from real experience. So many don't have actual experience. We don't have that many people who go out to the field, and then want to come back and teach; that's kind of a downer for them. The teaching today is very little in-between. They cram a 45 minute-class into 25 minutes. It's oversimplified, left out. A lot of it is, you've got one class that has several elements to talk about. They give them some materials, tell them to read it, and ask if there are any questions. Most students won't ask, they just want to move on. They're just checking off a box. Four or five years later, they come back for a re-take of the course. It's just a box check. Student quality is another issue. A lot of them can't write effectively at all. Some don't write, they are just on computers all the time. The instructors just laugh about it and move on. Some of them are high-ranked individuals, but they just do everything on the computer, they can't write! I did a lot of writing, that's actually what I did mostly as a contractor. It's just unbelievable. Some of them don't even know what team they are assigned to, you ask and they can't tell you. And these are people who were hired as intelligence officers. Some of them have other useful skills, others don't. They can't function without technology. It's sad.

When I enlisted for four years, it was just to give back to the country. Of course, it turned into twenty-four years in the Army. It got better, every time. Then when I retired, I hopped over to the Agency and started with SOG ▇▇▇▇▇▇▇. *Other than a few little bits, I never had it bad. I always got the jobs that I wanted.*

I enjoyed what I was doing. Now I'm getting to the point where I want to get out and do what WE want to do. To enjoy OUR lives. I would do it all over again, no problem. Nobody loves this country like I love this country.

My life has been an adventure. But it all just fell into place, one thing after another. I don't believe that I have been blessed. I've never really thought about it that way."

ACKNOWLEDGEMENTS:

Many people helped greatly in the production of this book. I am truly grateful to them all for their kind assistance. First and foremost, I owe my most heartfelt thanks to Dutch Wierenga. Thanks for letting me tell your incredible story, Dutch. I am proud to be your friend.

To the many members of MACV/SOG who generously provided me with their insights, images and corporate memories of those historic days. These include:

CSM JIM WHEELER, *1-0 of RT ALASKA, CCN. Jim, you were and are a wonderful role model for a young Agency officer. Mahalo, sir.*

SGM STEVE HOFFMAN, *1-1, RT ANACONDA, CCN, later a Squadron SGM of SFOD-D.*

MAJ JOHN PLASTER, *1-0, Covey Rider CCN/CCC, author*

SSG JOHN S. "TILT" MEYER, *1-0 of RT IDAHO, CCN, author*

MAJ HENRY KOHN, MACV/SOG, *Project SIGMA, First Reaction Company/Spike Team, CCS 1968-1969. Thanks for everything, Godfather.*

MSG DON MALEY, *1-0, RT ASP, CCN*

SGM CLIFF NEWMAN, *1-0, RT OHIO, CCN, Group Leader, First Combat HALO jump*

SFC MELVIN MCINTYRE, *CCS*

MR. JASON HARDY, *MACV/SOG Unit Historian*

MS. BONNIE COOPER, *Special Operations Association*

To all the men of MACV/SOG, living and deceased. Your deeds paved the way for the modern US Special Forces and CIA Paramilitary Officers, and those brave warriors stand on the shoulders of giants. America owes you all a tremendous debt of gratitude. You deserved far better than you received.

And finally, to my friends and colleagues at CIA, who stood the watch with me and extended me their respect and friendship whilst suffering together the slings and arrows of the Agency bureaucracy, and watching helplessly as it has descended into an enfeebled dysfunctional morass of "woke" political correctness which severely and negatively affects mission accomplishment. You know who you are. Thank you for your dedicated service to America. God bless you all.

The End.

"Kim Kipling" can be reached at SharpenedEdge@protonmail.com

ABOUT THE AUTHOR:

"Kim Kipling" (a pseudonym) is a retired U.S. Navy officer and CIA Paramilitary Operations Officer with over thirty years' service in CIA, both as a Staff Officer and a Contractor. He has known Dutch Wierenga, first as an instructor, then as a teammate and friend, for all that time. He has never been anywhere particularly interesting, nor done anything which might reasonably be considered noteworthy. He lives in Tennessee, with a three-legged tuxedo cat with the heart of a lion.

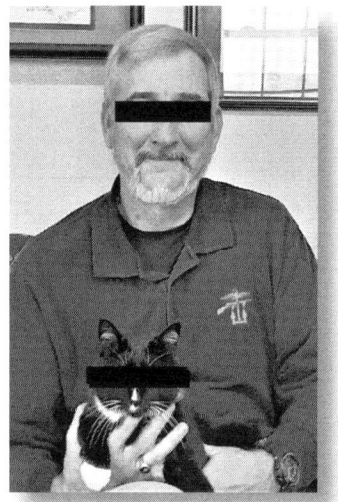

"KIM KIPLING"

INDEX:

A1E Skyraider, "Spad" – ch. 26
Bailey, Catherine A. – ch.18, 45
Bandung, Java – ch. 11
Bersiap, the – ch. 9
Boes, Helge, CIA – ch. 43
Bridge on the River Kwai – ch. 8
Burns, Michael P. SP4, MACV/SOG CCN – ch. 23
Byrd, Bradley – ch. 18
Byrd, Nathan – ch. 18
C-119 "Stinger" – ch. 26
Camp Mackall – ch. 31
Cedars, Robert SSG, RT ALASKA – ch. 23
Central Intelligence Agency -
    covert action – ch. 32
    founding, mission – ch. 32
    "The Farm" – ch. 35
    SOTC – ch. 37
    training – ch. 36 – 43
Changi Prison, Singapore – ch. 8
Chau Soc Rund, 0-1 RT ANACONDA – ch. 26, 27
Chau Thonh, RT ANACONDA – ch. 26, 27
Chau Penh, RT ANACONDA – ch. 26, 27
Contras, Nicaraguan – ch. 33
Covey Rider – ch. 24
Cronkite, Walter – ch. 20
"Death Railway" – ch. 8
"Demo Incident" – ch. 21
"Gorilla Incident" – ch. 21
Gonzalez, SFC, Bright Light mission – ch. 27
Groark, Thomas SP4, RT ALASKA – ch 23
Gulley, Robert SGT, RT ALASKA – ch. 24
Gurkhas – ch. 10, 20
HALO Committee – ch. 30
HALO parachuting – ch. 30

Hernandez, Sammy SFC, RT FLORIDA – ch. 30
Hill, Melvin SFC, RT FLORIDA – ch. 30
Hoffman, Steven SGM, RT ANACONDA– ch. 23, 25, 27-28
Hoang Cha Ly, RT ALASKA – ch. 23
Hopkins, SGM – ch. 23
Hornung, Ted SSG, MACV/SOG CCN – ch. 23, 25
Indonesian Civil War – ch. 9
Indonesian Internment Camp – ch. 9
Kalibeneng, Indonesia – ch. 4, 29
Kamai, Rocky SFC, RT ANACONDA – ch. 26
Kingbee, H-34 – ch. 24
Kim Teng, RT ANACONDA – ch. 26, 27
Kohn, Henry MAJ, MACV/SOG CCS, CIA – ch. 32
Lesesne, Ed MAJ MACV/SOG CCN – ch. 23
MACV/SOG
    history – ch. 22
    missions – ch 24.
    mission preparation – ch. 24
    personalities – ch. 25
    RT ANACONDA – ch. 26-28
    weapons – ch. 24
Magelang, Java – ch. 10
Mantracker School – ch. 20
McGuire Rig – ch. 24
Mcintyre, Melvin SFC, MACV/SOG CCS – ch. 21
Maley, Donald MSG, RT ANACONDA, RT ASP – ch 23, 25, 27
Meehan, Bob, COS – ch. 34
Miller, CPT, 2/7 SFG – ch. 31
Monroe, Jim, CIA SOG – ch. 32
Muntilan Internment Camp – ch. 8
Neal, Dennis P. 1LT – ch. 23
Newman, Cliff SGM, Bright Light mission, RT FLORIDA – ch. 25, 27, 30
Poelking, Jerry MAJ, CIA – ch. 32
"Phil" CIA, – ch. 32, 33
Raye, Martha COL – ch. 23
Rising Sun, Japan in WWII – ch. 5

RT ANACONDA
    air strike on horse – ch. 26
    air strikes on NVA base camp – ch. 26
    personnel – ch. 26
    wiretap mission – ch. 26
Rowe, James "Nick" COL – ch. 31
Schmutzer, Ignaz – ch. 11
SERE Program – ch. 31
STABO – ch. 24
Teeter, Roger SSG, RT ANACONDA – ch. 26
UH-1D "Huey" – ch. 24
Walton, John SP4, RT LOUISIANA – ch. 25
Warren, Jack COL, MACV/SOG CCN – ch. 23
Waugh, Billy SGM, MACV/SOG CCN, CIA – ch. 25
Wimmer, Kenneth R. SP4, RT ASP – ch. 23
Wheeler, Jim, CSM, RT ASP, RT ALASKA, CIA – ch. 23-25
Yend Day, RT ANACONDA – ch. 26, 27.
Wierenga, Jan W. "Dutch"
    birth – ch. 2
    CIA career – ch. 32-35
    CIA retirement – ch. 44
    combat patrols – ch. 12
    departure from Indonesia – ch. 14
    Dutch Army enlistment – ch. 15
    emigration to America – ch. 16
    family – ch. 2, 3
    first enlistment – ch. 15
    first marriage, immediate family – ch. 18
    imprisonment – ch. 8, 9
    MACV/SOG – ch. 23, 26-28
    Purple Heart – ch. 28
    Silver Star – ch. 27
    Special Forces training – ch 21
    today – ch. 46
    US Army basic training – ch. 17
Wierenga, Kathy D. – ch. 45

Made in the USA
Middletown, DE
07 May 2022